THERMODYNAMIC TABLES IN SI (METRIC) UNITS

(SYSTÈME INTERNATIONAL D'UNITÉS)

WITH CONVERSION FACTORS TO OTHER METRIC AND BRITISH UNITS

R. W. HAYWOOD

Fellow of St John's College and Reader in Engineering Thermodynamics in the University of Cambridge

SECOND EDITION

CAMBRIDGE UNIVERSITY PRESS

CAMBRIDGE

LONDON · NEW YORK · MELBOURNE

Published by the Syndics of the Cambridge University Press
The Pitt Building, Trumpington Street, Cambridge CB2 1RP
Bentley House, 200 Euston Road, London NW1 2DB
32 East 57th Street, New York, NY 10022, USA
296 Beaconsfield Parade, Middle Park, Melbourne 3206, Australia

ISBN 0 521 09714 2

First published 1968
Second edition 1972
Reprinted 1974 1976

Printed in Great Britain
at the
University Printing House, Cambridge
(Euan Phillips, University Printer)

PREFACE

These Tables in SI (metric) units have been modelled on the author's earlier Tables in British units, but further individual tables have been introduced. The steam tables have been extended to higher pressures and have been computed from more recent formulae. The table giving the properties of semi-perfect gases provides data for the calculation of enthalpies in the First Law analysis of combustion processes, while the tables of equilibrium constants and standard enthalpies of reaction cater for the effects of dissociation at higher temperatures. The tables giving the properties of air at low temperatures enable problems involving liquefaction and refrigeration at cryogenic temperatures to be handled, while the tables of transport properties of various fluids provide data for the solution of problems in heat transfer.

The most difficult decision in the construction of these new Tables arose in relation to the unit of pressure to be used. The basic SI unit of pressure, the N/m^2, is too small a unit for practical convenience in most engineering applications. As a result, the general tendency has been to favour the bar (10^5 N/m^2), which is about one atmosphere. However, 10^5 is not one of the recognised multiples of the *Système International*, which prefers steps of 10^3, and the use of multiples and submultiples of the bar would be even more objectionable. A still more serious objection to the use of the bar is that this perpetuates the need for a unit conversion factor in energy conversion calculations. Even though this factor is a multiple of 10, its presence disturbs the simple coherence of the *Système International*, which results from the fact that $1\,N = 1\,kg\,m/s^2$ and $1\,J = 1\,N\,m$. In a set of Tables designed primarily for use in Universities and Technical Colleges, it was consequently considered to be educationally desirable that tabulation should be primarily in terms of the N/m^2, and its recognised multiples; at the same time the corresponding number of bars has been indicated alongside each table at conveniently frequent intervals. In the saturation table for steam, it has been found convenient to change from the use of kN/m^2 to the use of MN/m^2 at about atmospheric pressure. In all other tables the pressure is quoted in MN/m^2. All pressures listed are **absolute** pressures.

To users of the Tables unfamiliar with SI units, attention is drawn to the fact that the kilogramme, not the gramme, is the basic SI unit of mass, and it is in terms of this unit that the SI is a coherent system of units, since $1\,N = 1\,kg\,m/s^2$. It is not practicable to express 10^{-3} kilogramme as a 'millikilogramme', so that it is still described as a 'gramme'. A similarly unsatisfactory situation exists in relation to the *kmol* and the *mol*. To counter this anomaly, a new name for the kilogramme mass is clearly needed; the author has suggested *berg* (unit symbol b), but international agreement on any change of name is not likely to be achieved in a short time.

The terminology of that branch of science concerned with temperature has long been in a most unsatisfactory state, and two Appendices are devoted to this topic in order to make clear the usage adopted in the Tables, in which all temperatures listed are thermodynamic in definition. In these Appendices, no mention is made of scale temperatures on the International Practical Scales of Temperature defined in 1948 and 1960, since a redefinition of these scales in terms of the kelvin unit of thermodynamic temperature is likely to be made in

1968.* It seems that there may then be good hope of achieving a rational simplification of nomenclature and symbology which will greatly reduce the current confusions arising from the present multiplicity of symbols. The usage adopted in the Tables is believed to be close to that which will eventually result from such simplification.

The stimulus for the production of these Tables was provided by the decision of the United Kingdom to commence the transition from British units to metric units of the *Système International*. Initially, many users of the Tables will be familiar with British units and with some commonly used non-SI metric units, but will be unfamiliar with SI units; with the passage of time, this situation will be reversed, but for many years engineers and scientists will need to consult earlier publications quoting data in British units and in non-SI metric units. The Appendices giving definitions and conversion factors have been prepared with this situation in mind. The inclusion of exact conversion factors is instructive in revealing the definitive relationships between the different systems of units. Rounded values for slide-rule calculation have also been quoted, but digital computers are increasingly being used and, for exact calculations, it is illogical to feed into such computers a conversion factor that has been rounded to the number of significant figures adequate only for slide-rule calculation. It would, of course, also be illogical to use exact conversion factors when such accuracy was not warranted. Throughout these Tables, exact numerical values are printed in **bold** type.

The author is especially grateful to all those on whose work he has drawn in compiling the Tables. He is also indebted to his colleagues for helpful advice in planning the initial lay-out of the Tables, to Dr J. H. Matthewman for programming the equations from which the property values for water substance and for Refrigerant-12 were computed and to Mr P. K. Clarkson for computational assistance in the preparation of various tables. Dr Matthewman and the author have prepared an enthalpy–entropy diagram for steam and a pressure–enthalpy diagram for Refrigerant-12; these were drawn by digital plotter from the same equations as those used in calculating the values for the respective tables and are available for use with the Tables.

R. W. HAYWOOD

Cambridge

* *The Rational Treatment of Temperature and Temperature Scales*, Haywood, R. W., *Proc. I. Mech. E.*, 1967–68, Volume 182, Part 1; contribution to the discussion by J. Terrien and J. de Boer. (This change was made in 1968. See *The International Practical Temperature Scale of 1968*, HMSO, London, 1969.)

SECOND EDITION 1972

Tables 7, 9, 10, 11 and 12 have been extended. The treatment of the *mole* has been brought into line with current international usage and a number of minor corrections and additions have been made.

CONTENTS

APPENDICES

GENERAL DATA

Precise data relating to temperature, definitions of units and unit conversion factors are given in the Appendices.

Temperature: At a *temperature* (thermodynamic) of T kelvins (namely T K), the corresponding truncated thermodynamic temperature called the *Celsius temperature* is *t Celsius*, written symbolically as t °C, where:

$$t = T - 273.15.$$

The difference in temperature between t_1 Celsius (t_1 °C) and t_2 Celsius (t_2 °C) is thus $(t_1 - t_2)$ kelvins, namely $(t_1 - t_2)$ K.

Note: Thermodynamic temperatures expressed in kelvins are commonly described as *absolute* temperatures.

Pressure:

1 bar $= 10^5$ N/m².

1 atm $=$ 1.01325 bar $=$ 0.101325 MN/m²
$=$ 760 torr $\approx$ 760 mmHg to 1 part in 7 million.

(For the definition of the mmHg, see Appendix C.)

Standard temperature and pressure (s.t.p.): 0 °C and 1 atm.

Atomic weights: Hydrogen 1*, Helium 4, Carbon 12, Nitrogen 14, Oxygen 16, Sulphur 32, Argon 40.

Mole: A *mole* (mol) is a unit of *quantity of particles of specified kind*; it is not a unit of mass. Its formal definition is given in Appendix A.

Molar mass: The *molar mass* is the mass in grammes (kilogrammes) of a mole (kilomole) of the specified substance.

Molar number (Avogadro constant): 6.023×10^{23} particles/mol ($\equiv 6.023 \times 10^{26}$ particles/kmol).

Electron charge: 1.60×10^{-19} coulomb.

Stefan–Boltzmann constant: 5.67×10^{-8} W/m² K⁴.

* A more exact value is 1.008.

TABLE 1. CALORIFIC VALUES

In this Table, the *calorific value* is the enthalpy decrease on combustion when the reactants and products are at 25 °C. In the evaluation of the lower calorific value the steam is taken as being dry saturated.

Substance	Molar mass kg/kmol	B.P. at 1 atm °C	Phase	Calorific value kJ/kg	
C (to CO)	12		sol.	9 190	
C (to CO_2)				32 760	
CO	28	−191.5	gas	10 100	
				Higher (gross) (H_2O to water)	Lower (net) (H_2O to steam)
H_2	2*	−252.9	gas	142 000	120 000
CH_4 (methane)	16	−161.5	gas	55 500	50 010
C_2H_6 (ethane)	30	−88.6	gas	51 870	47 470
C_3H_6 (propylene)	42	−47.7	gas	48 940	45 800
C_3H_8 (propane)	44	−42.1	gas	50 360	46 360
C_4H_{10} (butane)	58	−0.5	gas	49 520	45 730
C_8H_{18} (n-octane)	114	125.7	gas	48 270	44 800
			liq.	47 900	44 430

* A more exact value is 2.016.

PERFECT GASES

At normal atmospheric conditions, and over a limited range of temperature and pressure, the gases listed in Table 2 may be assumed to behave as perfect gases. That is, they may be assumed to have the equation of state $pv = RT$, and to have constant specific heats.

Molar (universal) gas constant: $\bar{R} = MR = 8.3143$ kJ/kmol K.

Molar volume of a perfect gas: 1 kmol of any perfect gas occupies a volume of approximately 22.4 m³ at s.t.p. (0 °C and 1 atm).

TABLE 2

Gas	Molar mass kg/kmol	Gas constant kJ/kg K	Specific heat capacity kJ/kg K c_p	c_v	c_p/c_v
Air	29.0	0.287	1.01	0.72	1.40
Atmospheric nitrogen†	28.15	0.295	1.03	0.74	1.40
N_2	28	0.297	1.04	0.74	1.40
O_2	32	0.260	0.92	0.66	1.40
A	40	0.208	0.52	0.31	1.67
H_2	2*	4.12	14.20	10.08	1.41
He	4	2.08	5.19	3.11	1.67
CO	28	0.297	1.04	0.74	1.40
CO_2	44	0.189	0.82	0.63	1.31
SO_2	64	0.130	0.61	0.48	1.26
CH_4	16	0.520	2.23	1.71	1.31
C_2H_6	30	0.277	1.75	1.47	1.19
C_3H_6	42	0.198	1.52	1.32	1.15

* A more exact value is 2.016.

† Air contains 0.93 % of argon (A) and traces of other gases; these and the nitrogen together are called *atmospheric nitrogen*.

Real gases are not perfect gases, and the rounded values for R, c_p, c_v and c_p/c_v listed above do not exactly satisfy the relationships between these quantities that would obtain for perfect gases.

Air composition:

Volumetric (and molar): 21.0 % O_2, 79.0 % atmospheric nitrogen.

Gravimetric: 23.2 % O_2, 76.8 % atmospheric nitrogen.

SEMI-PERFECT GASES

At low pressures, and over the temperature range quoted, the gases listed in this Table behave as semi-perfect gases. That is, while having the molar equation of state $p\bar{v} = \bar{R}T$, their specific heat capacities are variable but are functions only of temperature.

TABLE 3. MOLAR ENTHALPIES AT LOW PRESSURES

Warning: This table lists *absolute* temperatures.

Gas ...	Air	N_2	O_2	H_2	CO	CO_2	H_2O	
Molar mass / kg/kmol ...	29	28	32	2*	28	44	18	
Temperature / K	Molar enthalpy / MJ/kmol							Temperature / K
200	5.79	5.81	5.79	5.69	5.81	5.96	6.62	**200**
25 °C **298.15**	8.64	8.67	8.66	8.46	8.67	9.37	9.90	**298.15** 25 °C
300	8.70	8.72	8.71	8.52	8.72	9.44	9.96	**300**
400	11.62	11.64	11.68	11.42	11.64	13.37	13.35	**400**
500	14.57	14.58	14.74	14.34	14.60	17.67	16.82	**500**
600	17.59	17.56	17.90	17.27	17.61	22.27	20.39	**600**
700	20.66	20.61	21.16	20.21	20.69	27.12	24.09	**700**
800	23.81	23.72	24.50	23.16	23.85	32.18	27.90	**800**
900	27.03	26.89	27.90	26.13	27.07	37.41	31.83	**900**
1000	30.30	30.14	31.37	29.14	30.36	42.78	35.90	**1000**
1100	33.64	33.44	34.88	32.18	33.71	48.27	40.09	**1100**
1200	37.02	36.79	38.43	35.26	37.11	53.87	44.41	**1200**
1300	40.44	40.19	42.01	38.38	40.54	59.55	48.84	**1300**
1400	43.90	43.62	45.63	41.54	44.02	65.31	53.39	**1400**
1500	47.39	47.09	49.27	44.75	47.53	71.13	58.05	**1500**
1600	50.92	50.59	52.94	48.00	51.07	77.01	62.81	**1600**
1700	54.47	54.12	56.63	51.29	54.63	82.94	67.65	**1700**
1800	58.04	57.67	60.35	54.62	58.21	88.92	72.58	**1800**
1900	61.63	61.25	64.09	58.00	61.81	94.93	77.59	**1900**
2000	65.24	64.84	67.86	61.40	65.42	100.97	82.67	**2000**
2100	68.87	68.44	71.65	64.84	69.06	107.05	87.81	**2100**
2200	72.52	72.06	75.46	68.31	72.70	113.15	93.01	**2200**
2300	76.18	75.70	79.29	71.82	76.36	119.28	98.27	**2300**
2400	79.86	79.35	83.14	75.35	80.03	125.43	103.58	**2400**
2500	83.55	83.01	87.02	78.90	83.71	131.61	108.94	**2500**
2600	87.25	86.68	90.92	82.48	87.40	137.80	114.34	**2600**
2700	90.96	90.36	94.83	86.09	91.10	144.02	119.78	**2700**
2800	94.69	94.05	98.77	89.72	94.80	150.25	125.26	**2800**
2900	98.42	97.74	102.72	93.37	98.51	156.50	130.77	**2900**
3000	102.16	101.44	106.70	97.04	102.23	162.76	136.31	**3000**

* A more exact value is 2.016.

Note: (1) The molar enthalpies listed are those in the ideal gas state at zero pressure, but the values given are also valid at and around atmospheric pressure.

(2) In this table, the arbitrary datum state for zero enthalpy is that of the substance in the ideal gas state at zero pressure and zero absolute temperature. (**Warning**: In Tables 6–12, the arbitrary datum state for H_2O is that of the saturated liquid at the triple point, at which state the internal energy and entropy are taken to be zero.)

THERMOCHEMICAL DATA FOR EQUILIBRIUM REACTIONS

TABLES 4 AND 5 RELATE TO THE REACTIONS LISTED BELOW

STOICHIOMETRIC EQUATIONS

$$\Sigma \nu_i A_i = 0,$$

where ν_i is the *stoichiometric coefficient* of the substance whose *chemical symbol* is A_i.

(1) $-2H + H_2 = 0$

(2) $-2N + N_2 = 0$

(3) $-2O + O_2 = 0$

(4) $-2NO + N_2 + O_2 = 0$

(5) $-H_2 - \frac{1}{2}O_2 + H_2O = 0$

(6) $-\frac{1}{2}H_2 - OH + H_2O = 0$

(7) $-CO - \frac{1}{2}O_2 + CO_2 = 0$

(8) $-CO - H_2O + CO_2 + H_2 = 0$

(9) $-\frac{1}{2}N_2 - \frac{3}{2}H_2 + NH_3 = 0$

EQUILIBRIUM CONSTANTS

$$\log_{10} K_p = \sum_i \nu_i \log_{10} p'_i,$$

where p'_i = partial pressure of substance A_i, in atmospheres (atm).

TABLE 4. EQUILIBRIUM CONSTANTS

Warning: This table lists *absolute* temperatures.

Temp. K	Reaction number 1	2	3	4	5	6	7	8	9	Temp. K
	$\log_{10} K_p$									
200	108.644	240.810	123.984	45.858	60.792	70.267	69.356	8.564	6.708	**200**
298	71.224	159.600	81.208	30.342	40.048	46.137	45.066	5.018	2.869	**298**
400	51.752	117.408	58.946	22.284	29.240	33.567	32.431	3.191	0.778	**400**
600	32.672	76.162	37.148	14.420	18.633	21.242	20.087	1.454	−1.380	**600**
800	23.078	55.488	26.202	10.486	13.289	15.044	13.916	0.627	−2.523	**800**
1000	17.292	43.056	19.614	8.124	10.062	11.309	10.221	0.159	−3.233	**1000**
1200	13.414	34.754	15.208	6.550	7.899	8.811	7.764	−0.135	−3.716	**1200**
1400	10.630	28.812	12.054	5.424	6.347	7.021	6.014	−0.333	−4.064	**1400**
1600	8.532	24.350	9.684	4.580	5.180	5.677	4.706	−0.474	−4.325	**1600**
1800	6.896	20.874	7.836	3.924	4.270	4.631	3.693	−0.577	−4.528	**1800**
2000	5.580	18.092	6.356	3.398	3.540	3.793	2.884	−0.656	−4.689	**2000**
2200	4.502	15.810	5.142	2.968	2.942	3.107	2.226	−0.716	−4.819	**2200**
2400	3.600	13.908	4.130	2.610	2.443	2.535	1.679	−0.764	−4.927	**2400**
2600	2.834	12.298	3.272	2.308	2.021	2.052	1.219	−0.802	−5.016	**2600**
2800	2.178	10.914	2.536	2.050	1.658	1.637	0.825	−0.833	−5.092	**2800**
3000	1.606	9.716	1.898	1.826	1.343	1.278	0.485	−0.858	−5.156	**3000**
3200	1.106	8.664	1.340	1.630	1.067	0.963	0.189	−0.878	−5.211	**3200**
3400	0.664	7.736	0.846	1.458	0.824	0.687	−0.071	−0.895	−5.259	**3400**
3600	0.270	6.910	0.408	1.306	0.607	0.440	−0.302	−0.909	−5.300	**3600**
3800	−0.084	6.172	0.014	1.170	0.413	0.220	−0.508	−0.921	−5.336	**3800**
4000	−0.402	5.504	−0.340	1.048	0.238	0.022	−0.692	−0.930	−5.368	**4000**
4500	−1.074	4.094	−1.086	0.794	−0.133	−0.397	−1.079	−0.946	−5.431	**4500**
5000	−1.612	2.962	−1.686	0.592	−0.430	−0.731	−1.386	−0.956	−5.477	**5000**
5500	−2.054	2.032	−2.176	0.428	−0.675	−1.004	−1.635	−0.960	−5.511	**5500**
6000	−2.422	1.250	−2.584	0.294	−0.880	−1.232	−1.841	−0.961	−5.536	**6000**

STANDARD FREE ENTHALPY OF REACTION

At a given temperature, the standard free enthalpy of reaction ΔG°_T (or *standard Gibbs function change*) may be calculated from the listed value of $\log_{10} K_p$ by the following equation:

$$\Delta G^\circ_T = -\bar{R}T \log_e K_p$$

$$= -0.01914\, T \log_{10} K_p \text{ MJ.}$$

STANDARD ENTHALPY OF REACTION

$$\Delta H_T^\circ = \sum_i \nu_i [\bar{h}_i]_T^\circ,$$

where $[\bar{h}_i]_T^\circ$ = enthalpy per kmol of substance A_i, at 1 atm pressure and absolute temperature T.

TABLE 5. STANDARD ENTHALPY OF REACTION

Warning: This table lists *absolute* temperatures.

Temp. K	Reaction number 1	2	3	4	5	6	7	8	9	Temp. K
	ΔH_T°/MJ									
200	−434.7	−944.1	−496.9	−180.4	−240.9	−280.2	−282.1	−41.21	−43.71	**200**
298	−436.0	−945.3	−498.4	−180.6	−241.8	−281.3	−283.0	−41.17	−45.90	**298**
400	−437.3	−946.6	−499.8	−180.7	−242.8	−282.4	−283.5	−40.63	−48.04	**400**
600	−439.7	−948.9	−502.1	−180.7	−244.8	−284.1	−283.6	−38.88	−51.39	**600**
800	−442.1	−951.1	−503.9	−180.8	−246.5	−285.5	−283.3	−36.82	−53.66	**800**
1000	−444.5	−953.0	−505.4	−180.9	−247.9	−286.6	−282.6	−34.74	−55.07	**1000**
1200	−446.7	−954.7	−506.7	−180.9	−249.0	−287.4	−281.8	−32.79	−55.83	**1200**
1400	−448.7	−956.1	−507.8	−181.0	−249.9	−287.9	−280.9	−30.98	−56.07	**1400**
1600	−450.6	−957.5	−508.9	−181.0	−250.6	−288.4	−279.9	−29.29	−55.99	**1600**
1800	−452.3	−958.7	−509.8	−181.0	−251.2	−288.6	−278.9	−27.71	−55.66	**1800**
2000	−453.8	−959.9	−510.6	−181.0	−251.7	−288.8	−277.9	−26.22	−55.19	**2000**
2200	−455.2	−961.0	−511.4	−180.8	−252.1	−288.9	−276.8	−24.79	−54.61	**2200**
2400	−456.4	−962.1	−512.0	−180.7	−252.4	−289.0	−275.8	−23.41	−53.92	**2400**
2600	−457.6	−963.1	−512.5	−180.4	−252.7	−289.0	−274.8	−22.07	−53.12	**2600**
2800	−458.6	−964.1	−513.0	−180.1	−253.0	−288.9	−273.7	−20.77	−52.22	**2800**
3000	−459.6	−965.0	−513.4	−179.7	−253.3	−288.9	−272.7	−19.49	−51.20	**3000**
3200	−460.4	−966.0	−513.8	−179.3	−253.5	−288.8	−271.7	−18.19	−50.10	**3200**
3400	−461.2	−967.0	−514.1	−178.7	−253.8	−288.7	−270.7	−16.91	−48.94	**3400**
3600	−461.9	−968.1	−514.4	−178.2	−254.1	−288.6	−269.8	−15.62	−47.75	**3600**
3800	−462.5	−969.2	−514.6	−177.6	−254.5	−288.5	−268.8	−14.33	−46.49	**3800**
4000	−463.0	−970.4	−514.8	−176.9	−254.8	−288.4	−267.8	−13.00	−45.19	**4000**
4500	−464.0	−973.8	−515.3	−175.2	−255.9	−288.1	−265.5	−9.57	−41.68	**4500**
5000	−464.6	−977.9	−515.9	−173.2	−257.2	−288.0	−263.1	−5.95	−37.79	**5000**
5500	−464.8	−982.9	−516.5	−171.1	−258.6	−287.9	−260.7	−2.10	−33.56	**5500**
6000	−464.7	−989.0	−517.2	−169.0	−260.3	−287.9	−258.2	2.01	−28.98	**6000**

STEAM TABLES

In Tables 6–12, giving the thermodynamic properties of ordinary water substance, the arbitrary datum state for zero internal energy and entropy is that of the saturated liquid at the triple point, which is the state point at which the solid, liquid and vapour are together in equilibrium.

TABLE 6. TRIPLE POINT OF WATER

TEMPERATURE: **273.16** K
CELSIUS TEMPERATURE: **0.01** °C
PRESSURE: 0.6112 kN/m^2

Phase	Specific volume m^3/kg	Specific enthalpy kJ/kg	Specific entropy kJ/kg K
Ice	1.0905×10^{-3}	−333.5	−1.221
Water	1.0002×10^{-3}	0.000 61	**zero**
Steam	206.2	2501.6	9.157

TABLE 7. SATURATED WATER AND STEAM

TEMPERATURES FROM THE TRIPLE POINT TO **100** °C

[**100** kN/m² = **1** bar ≈ 14.5 lbf/in²]

Celsius temp., °C	Pressure kN/m²	Specific volume m³/kg		Specific internal energy kJ/kg		Specific enthalpy kJ/kg			Specific entropy kJ/kg K		Celsius temp., °C
		Water	Steam	Water	Steam	Water	Evaporation	Steam	Water	Steam	
t	p	v_f	v_g	u_f	u_g	h_f	h_{fg}	h_g	s_f	s_g	t
0.01	0.611	0.001000	206.2	**zero**	2375.6	+0.0	2501.6	2501.6	**zero**	9.157	**0.01 Triple point**
2	0.705	0.001000	179.9	8.4	2378.3	8.4	2496.8	2505.2	0.031	9.105	**2**
4	0.813	0.001000	157.3	16.8	2381.1	16.8	2492.1	2508.9	0.061	9.053	**4**
6	0.935	0.001000	137.8	25.2	2383.8	25.2	2487.4	2512.6	0.091	9.001	**6**
8	1.072	0.001000	121.0	33.6	2386.6	33.6	2482.6	2516.2	0.121	8.951	**8**
10	1.227	0.001000	106.4	42.0	2389.3	42.0	2477.9	2519.9	0.151	8.902	**10**
12	1.401	0.001000	93.8	50.4	2392.1	50.4	2473.2	2523.6	0.180	8.854	**12**
14	1.597	0.001001	82.9	58.8	2394.8	58.8	2468.5	2527.2	0.210	8.806	**14**
16	1.817	0.001001	73.4	67.1	2397.6	67.1	2463.8	2530.9	0.239	8.759	**16**
18	2.062	0.001001	65.1	75.5	2400.3	75.5	2459.0	2534.5	0.268	8.713	**18**
20	2.34	0.001002	57.8	83.9	2403.0	83.9	2454.3	2538.2	0.296	8.668	**20**
22	2.64	0.001002	51.5	92.2	2405.8	92.2	2449.6	2541.8	0.325	8.624	**22**
24	2.98	0.001003	45.9	100.6	2408.5	100.6	2444.9	2545.5	0.353	8.581	**24**
25	3.17	0.001003	43.4	104.8	2409.9	104.8	2442.5	2547.3	0.367	8.559	**25**
26	3.36	0.001003	41.0	108.9	2411.2	108.9	2440.2	2549.1	0.381	8.538	**26**
28	3.78	0.001004	36.7	117.3	2414.0	117.3	2435.4	2552.7	0.409	8.496	**28**
30	4.24	0.001004	32.9	125.7	2416.7	125.7	2430.7	2556.4	0.437	8.455	**30**
32	4.75	0.001005	29.6	134.0	2419.4	134.0	2425.9	2560.0	0.464	8.414	**32**
34	5.32	0.001006	26.6	142.4	2422.1	142.4	2421.2	2563.6	0.491	8.374	**34**
36	5.94	0.001006	24.0	150.7	2424.8	150.7	2416.4	2567.2	0.518	8.335	**36**
38	6.62	0.001007	21.6	159.1	2427.5	159.1	2411.7	2570.8	0.545	8.296	**38**
40	7.38	0.001008	19.55	167.4	2430.2	167.5	2406.9	2574.4	0.572	8.258	**40**
42	8.20	0.001009	17.69	175.8	2432.9	175.8	2402.1	2577.9	0.599	8.221	**42**
44	9.10	0.001009	16.04	184.2	2435.6	184.2	2397.3	2581.5	0.625	8.184	**44**
46	10.09	0.001010	14.56	192.5	2438.3	192.5	2392.5	2585.1	0.651	8.148	**46**
48	11.16	0.001011	13.23	200.9	2440.9	200.9	2387.7	2588.6	0.678	8.113	**48**

t	p	v_f	v_g	u_f	u_g	h_f	h_{fg}	h_g	s_f	s_g	t
50	12.34	0.001012	12.05	209.2	2443.6	209.3	2382.9	2592.2	0.704	8.078	**50**
52	13.61	0.001013	10.98	217.6	2446.2	217.6	2378.1	2595.7	0.729	8.043	**52**
54	15.00	0.001014	10.02	226.0	2448.9	226.0	2373.2	2599.2	0.755	8.009	**54**
56	16.51	0.001015	9.16	234.3	2451.5	234.4	2368.4	2602.7	0.780	7.976	**56**
58	18.15	0.001016	8.38	242.7	2454.1	242.7	2363.5	2606.2	0.806	7.943	**58**
60	19.92	0.001017	7.68	251.1	2456.8	251.1	2358.6	2609.7	0.831	7.911	**60**
62	21.84	0.001018	7.04	259.4	2459.4	259.5	2353.7	2613.2	0.856	7.879	**62**
64	23.91	0.001019	6.47	267.8	2462.0	267.8	2348.8	2616.6	0.881	7.848	**64**
66	26.15	0.001020	5.95	276.2	2464.5	276.2	2343.9	2620.1	0.906	7.817	**66**
68	28.56	0.001022	5.48	284.6	2467.1	284.6	2338.9	2623.5	0.930	7.786	**68**
70	31.16	0.001023	5.05	292.9	2469.7	293.0	2334.0	2626.9	0.955	7.756	**70**
72	33.96	0.001024	4.66	301.3	2472.2	301.4	2329.0	2630.3	0.979	7.727	**72**
74	36.96	0.001025	4.30	309.7	2474.8	309.7	2324.0	2633.7	1.003	7.698	**74**
76	40.19	0.001027	3.98	318.1	2477.3	318.1	2318.9	2637.1	1.027	7.669	**76**
78	43.65	0.001028	3.68	326.5	2479.8	326.5	2313.9	2640.4	1.051	7.641	**78**
80	47.36	0.001029	3.41	334.9	2482.3	334.9	2308.8	2643.8	1.075	7.613	**80**
82	51.33	0.001031	3.16	343.3	2484.8	343.3	2303.8	2647.1	1.099	7.586	**82**
84	55.57	0.001032	2.93	351.7	2487.3	351.7	2298.6	2650.4	1.123	7.559	**84**
86	60.11	0.001033	2.73	360.1	2489.7	360.1	2293.5	2653.6	1.146	7.532	**86**
88	64.95	0.001035	2.54	368.5	2492.2	368.5	2288.4	2656.9	1.169	7.506	**88**
90	70.11	0.001036	2.36	376.9	2494.6	376.9	2283.2	2660.1	1.193	7.480	**90**
92	75.61	0.001038	2.20	385.3	2497.0	385.4	2278.0	2663.4	1.216	7.454	**92**
94	81.46	0.001039	2.05	393.7	2499.4	393.8	2272.8	2666.6	1.239	7.429	**94**
96	87.69	0.001041	1.915	402.1	2501.8	402.2	2267.5	2669.7	1.261	7.404	**96**
98	94.30	0.001042	1.789	410.5	2504.1	410.6	2262.2	2672.9	1.284	7.380	**98**
100	101.325	0.001044	1.673	419.0	2506.5	419.1	2256.9	2676.0	1.307	7.355	**100**

TABLE 8. SATURATED WATER AND STEAM

PRESSURES FROM THE TRIPLE POINT UP TO 100 kN/m² (1 BAR)

[100 kN/m² = 1 bar ≈ 14.5 lbf/in²]

Pressure kN/m²	Celsius temp., °C	Specific volume m³/kg		Specific internal energy kJ/kg		Specific enthalpy kJ/kg			Specific entropy kJ/kg K		Pressure kN/m²	
		Water	Steam	Water	Steam	Water	Evaporation	Steam	Water	Steam		
p	t	v_f	v_g	u_f	u_g	h_f	h_{fg}	h_g	s_f	s_g	p	
0.611	**0.01**	0.001000	206.2	**zero**	2375.6	+0.0	2501.6	2501.6	**zero**	9.157	0.611	**Triple point**
0.8	3.8	0.001000	159.7	15.8	2380.7	15.8	2492.6	2508.5	0.058	9.058	**0.8**	
1.0	7.0	0.001000	129.2	29.3	2385.2	29.3	2485.0	2514.4	0.106	8.977	**1.0**	**0.01** bar
1.2	9.7	0.001000	108.7	40.6	2388.9	40.6	2478.7	2519.3	0.146	8.910	**1.2**	
1.4	12.0	0.001000	93.9	50.3	2392.0	50.3	2473.2	2523.5	0.180	8.854	**1.4**	
1.6	14.0	0.001001	82.8	58.9	2394.8	58.9	2468.4	2527.3	0.210	8.805	**1.6**	
1.8	15.9	0.001001	74.0	66.5	2397.4	66.5	2464.1	2530.6	0.237	8.763	**1.8**	
2.0	17.5	0.001001	67.0	73.5	2399.6	73.5	2460.2	2533.6	0.261	8.725	**2.0**	
2.2	19.0	0.001002	61.2	79.8	2401.7	79.8	2456.6	2536.4	0.282	8.690	**2.2**	
2.4	20.4	0.001002	56.4	85.7	2403.6	85.7	2453.3	2539.0	0.302	8.659	**2.4**	
2.6	21.7	0.001002	52.3	91.1	2405.4	91.1	2450.2	2541.3	0.321	8.630	**2.6**	
2.8	23.0	0.001002	48.7	96.2	2407.1	96.2	2447.3	2543.6	0.338	8.603	**2.8**	
3.0	24.1	0.001003	45.7	101.0	2408.6	101.0	2444.6	2545.6	0.354	8.578	**3.0**	
3.5	26.7	0.001003	39.5	111.8	2412.2	111.8	2438.5	2550.4	0.391	8.523	**3.5**	
4.0	29.0	0.001004	34.8	121.4	2415.3	121.4	2433.1	2554.5	0.422	8.475	**4.0**	
4.5	31.0	0.001005	31.1	130.0	2418.1	130.0	2428.2	2558.2	0.451	8.433	**4.5**	
5.0	32.9	0.001005	28.2	137.8	2420.6	137.8	2423.8	2561.6	0.476	8.396	**5.0**	
6	36.2	0.001006	23.74	151.5	2425.1	151.5	2416.0	2567.5	0.521	8.331	**6**	
7	39.0	0.001007	20.53	163.4	2428.9	163.4	2409.2	2572.6	0.559	8.277	**7**	
8	41.5	0.001008	18.10	173.9	2432.3	173.9	2403.2	2577.1	0.593	8.230	**8**	
9	43.8	0.001009	16.20	183.3	2435.3	183.3	2397.9	2581.1	0.622	8.188	**9**	
10	45.8	0.001010	14.67	191.8	2438.0	191.8	2392.9	2584.8	0.649	8.151	**10**	**0.1** bar
11	47.7	0.001011	13.42	199.7	2440.5	199.7	2388.4	2588.1	0.674	8.118	**11**	
12	49.4	0.001012	12.36	206.9	2442.8	206.9	2384.3	2591.2	0.696	8.087	**12**	
13	51.1	0.001013	11.47	213.7	2445.0	213.7	2380.4	2594.0	0.717	8.059	**13**	
14	52.6	0.001013	10.69	220.0	2447.0	220.0	2376.7	2596.7	0.737	8.033	**14**	
15	54.0	0.001014	10.02	226.0	2448.9	226.0	2373.2	2599.2	0.755	8.009	**15**	

Pressure in **kN/m²**

p	t	v_f	v_g	u_f	u_g	h_f	h_{fg}	h_g	s_f	s_g	p	
16	55.3	0.001015	9.43	231.6	2450.6	231.6	2370.0	2601.6	0.772	7.987	**16**	
17	56.6	0.001015	8.91	236.9	2452.3	236.9	2366.9	2603.8	0.788	7.966	**17**	
18	57.8	0.001016	8.45	242.0	2453.9	242.0	2363.9	2605.9	0.804	7.946	**18**	
19	59.0	0.001017	8.03	246.8	2455.4	246.8	2361.1	2607.9	0.818	7.927	**19**	
20	60.1	0.001017	7.65	251.4	2456.9	251.5	2358.4	2609.9	0.832	7.909	**20**	**0.2** bar
22	62.2	0.001018	7.00	260.1	2459.6	260.1	2353.3	2613.5	0.858	7.876	**22**	
24	64.1	0.001019	6.45	268.2	2462.1	268.2	2348.6	2616.8	0.882	7.846	**24**	
26	65.9	0.001020	5.98	275.6	2464.4	275.7	2344.2	2619.9	0.904	7.819	**26**	
28	67.5	0.001021	5.58	282.7	2466.5	282.7	2340.0	2622.7	0.925	7.793	**28**	
30	69.1	0.001022	5.23	289.3	2468.6	289.3	2336.1	2625.4	0.944	7.770	**30**	
35	72.7	0.001025	4.53	304.3	2473.1	304.3	2327.2	2631.5	0.988	7.717	**35**	
40	75.9	0.001027	3.99	317.6	2477.1	317.7	2319.2	2636.9	1.026	7.671	**40**	
45	78.7	0.001028	3.58	329.6	2480.7	329.6	2312.0	2641.7	1.060	7.631	**45**	
50	81.3	0.001030	3.24	340.5	2484.0	340.6	2305.4	2646.0	1.091	7.595	**50**	**0.5** bar
55	83.7	0.001032	2.96	350.6	2486.9	350.6	2299.3	2649.9	1.119	7.562	**55**	
60	86.0	0.001033	2.73	359.9	2489.7	359.9	2293.6	2653.6	1.145	7.533	**60**	
65	88.0	0.001035	2.53	368.5	2492.2	368.6	2288.3	2656.9	1.170	7.506	**65**	
70	90.0	0.001036	2.36	376.7	2494.5	376.8	2283.3	2660.1	1.192	7.480	**70**	
75	91.8	0.001037	2.22	384.4	2496.7	384.5	2278.6	2663.0	1.213	7.457	**75**	
80	93.5	0.001039	2.087	391.6	2498.8	391.7	2274.1	2665.8	1.233	7.435	**80**	
85	95.2	0.001040	1.972	398.5	2500.8	398.6	2269.8	2668.4	1.252	7.415	**85**	
90	96.7	0.001041	1.869	405.1	2502.6	405.2	2265.6	2670.9	1.270	7.395	**90**	
95	98.2	0.001042	1.777	411.4	2504.4	411.5	2261.7	2673.2	1.287	7.377	**95**	
100	99.6	0.001043	1.694	417.4	2506.1	417.5	2257.9	2675.4	1.303	7.360	**100**	**1** bar
101.325	100.0	0.001044	1.673	419.0	2506.5	419.1	2256.9	2676.0	1.307	7.355	**101.325**	**1** atm

TABLE 8 (*cont.*). SATURATED WATER AND STEAM

PRESSURES FROM **0.1** TO **3.0** MN/m² (**1** TO **30** BAR)

[0.1 MN/m² = 1 bar ≈ 14.5 lbf/in²]

Pressure MN/m² p	Celsius temp., °C t	Specific volume m³/kg Water v_f	Steam v_g	Specific internal energy kJ/kg Water u_f	Steam u_g	Specific enthalpy kJ/kg Water h_f	Evaporation h_{fg}	Steam h_g	Specific entropy kJ/kg K Water s_f	Steam s_g	Pressure MN/m² p	
0.10	99.6	0.001043	1.694	417.4	2506.1	417.5	2257.9	2675.4	1.303	7.360	**0.10**	**1** bar
0.11	102.3	0.001046	1.549	428.7	2509.2	428.8	2250.8	2679.6	1.333	7.328	**0.11**	
0.12	104.8	0.001048	1.428	439.2	2512.1	439.4	2244.1	2683.4	1.361	7.298	**0.12**	
0.13	107.1	0.001049	1.325	449.1	2514.7	449.2	2237.8	2687.0	1.387	7.271	**0.13**	
0.14	109.3	0.001051	1.236	458.3	2517.2	458.4	2231.9	2690.3	1.411	7.247	**0.14**	
0.15	111.4	0.001053	1.159	467.0	2519.5	467.1	2226.2	2693.4	1.434	7.223	**0.15**	
0.16	113.3	0.001055	1.091	475.2	2521.7	475.4	2220.9	2696.2	1.455	7.202	**0.16**	
0.17	115.2	0.001056	1.031	483.0	2523.7	483.2	2215.7	2699.0	1.475	7.181	**0.17**	
0.18	116.9	0.001058	0.977	490.5	2525.6	490.7	2210.8	2701.5	1.494	7.162	**0.18**	
0.19	118.6	0.001059	0.929	497.6	2527.5	497.8	2206.1	2704.0	1.513	7.144	**0.19**	
0.20	120.2	0.001061	0.885	504.5	2529.2	504.7	2201.6	2706.3	1.530	7.127	**0.20**	**2** bar
0.22	123.3	0.001064	0.810	517.4	2532.4	517.6	2193.0	2710.6	1.563	7.095	**0.22**	
0.24	126.1	0.001066	0.746	529.4	2535.4	529.6	2184.9	2714.5	1.593	7.066	**0.24**	
0.26	128.7	0.001069	0.693	540.6	2538.1	540.9	2177.3	2718.2	1.621	7.039	**0.26**	
0.28	131.2	0.001071	0.646	551.1	2540.6	551.4	2170.1	2721.5	1.647	7.014	**0.28**	
0.30	133.5	0.001074	0.606	561.1	2543.0	561.4	2163.2	2724.7	1.672	6.991	**0.30**	
0.32	135.8	0.001076	0.570	570.6	2545.2	570.9	2156.7	2727.6	1.695	6.969	**0.32**	
0.34	137.9	0.001078	0.538	579.6	2547.2	579.9	2150.4	2730.3	1.717	6.949	**0.34**	
0.36	139.9	0.001080	0.510	588.1	2549.2	588.5	2144.4	2732.9	1.738	6.930	**0.36**	
0.38	141.8	0.001082	0.485	596.4	2551.0	596.8	2138.6	2735.3	1.757	6.912	**0.38**	
0.40	143.6	0.001084	0.462	604.2	2552.7	604.7	2133.0	2737.6	1.776	6.894	**0.40**	**4** bar
0.42	145.4	0.001086	0.442	611.8	2554.4	612.3	2127.5	2739.8	1.795	6.878	**0.42**	
0.44	147.1	0.001088	0.423	619.1	2555.9	619.6	2122.3	2741.9	1.812	6.862	**0.44**	
0.46	148.7	0.001089	0.405	626.2	2557.4	626.7	2117.2	2743.9	1.829	6.847	**0.46**	
0.48	150.3	0.001091	0.389	633.0	2558.8	633.5	2112.2	2745.7	1.845	6.833	**0.48**	
0.50	151.8	0.001093	0.375	639.6	2560.2	640.1	2107.4	2747.5	1.860	6.819	**0.50**	
0.55	155.5	0.001097	0.342	655.2	2563.3	655.8	2095.9	2751.7	1.897	6.787	**0.55**	
0.60	158.8	0.001101	0.315	669.8	2566.2	670.4	2085.0	2755.5	1.931	6.758	**0.60**	
0.65	162.0	0.001105	0.292	683.4	2568.7	684.1	2074.7	2758.9	1.962	6.730	**0.65**	
0.70	165.0	0.001108	0.273	696.3	2571.1	697.1	2064.9	2762.0	1.992	6.705	**0.70**	**7** bar

Pressure in **MN/m²**

p	t	v_f	v_g	u_f	u_g	h_f	h_{fg}	h_g	s_f	s_g	p	
0.75	167.8	0.001112	0.2554	708.5	2573.3	709.3	2055.5	2764.8	2.020	6.682	**0.75**	
0.80	170.4	0.001115	0.2403	720.0	2575.3	720.9	2046.5	2767.5	2.046	6.660	**0.80**	
0.85	172.9	0.001118	0.2268	731.1	2577.1	732.0	2037.9	2769.9	2.071	6.639	**0.85**	
0.90	175.4	0.001121	0.2148	741.6	2578.8	742.6	2029.5	2772.1	2.094	6.619	**0.90**	
0.95	177.7	0.001124	0.2040	751.8	2580.4	752.8	2021.4	2774.2	2.117	6.601	**0.95**	
1.00	179.9	0.001127	0.1943	761.5	2581.9	762.6	2013.6	2776.2	2.138	6.583	**1.00**	**10** bar
1.05	182.0	0.001130	0.1855	770.8	2583.3	772.0	2005.9	2778.0	2.159	6.566	**1.05**	
1.10	184.1	0.001133	0.1774	779.9	2584.5	781.1	1998.5	2779.7	2.179	6.550	**1.10**	
1.15	186.0	0.001136	0.1700	788.6	2585.8	789.9	1991.3	2781.3	2.198	6.534	**1.15**	
1.20	188.0	0.001139	0.1632	797.1	2586.9	798.4	1984.3	2782.7	2.216	6.519	**1.20**	
1.25	189.8	0.001141	0.1569	805.3	2588.0	806.7	1977.4	2784.1	2.234	6.505	**1.25**	
1.30	191.6	0.001144	0.1511	813.2	2589.0	814.7	1970.7	2785.4	2.251	6.491	**1.30**	
1.4	195.0	0.001149	0.1407	828.5	2590.8	830.1	1957.7	2787.8	2.284	6.465	**1.4**	
1.5	198.3	0.001154	0.1317	842.9	2592.4	844.7	1945.2	2789.9	2.314	6.441	**1.5**	
1.6	201.4	0.001159	0.1237	856.7	2593.8	858.6	1933.2	2791.7	2.344	6.418	**1.6**	
1.7	204.3	0.001163	0.1166	869.9	2595.1	871.8	1921.5	2793.4	2.371	6.396	**1.7**	
1.8	207.1	0.001168	0.1103	882.5	2596.3	884.6	1910.3	2794.8	2.398	6.375	**1.8**	
1.9	209.8	0.001172	0.1047	894.6	2597.3	896.8	1899.3	2796.1	2.423	6.355	**1.9**	
2.0	212.4	0.001177	0.0995	906.2	2598.2	908.6	1888.6	2797.2	2.447	6.337	**2.0**	**20** bar
2.1	214.9	0.001181	0.0949	917.5	2598.9	920.0	1878.2	2798.2	2.470	6.319	**2.1**	
2.2	217.2	0.001185	0.0907	928.3	2599.6	931.0	1868.1	2799.1	2.492	6.301	**2.2**	
2.3	219.6	0.001189	0.0868	938.9	2600.2	941.6	1858.2	2799.8	2.514	6.285	**2.3**	
2.4	221.8	0.001193	0.0832	949.1	2600.7	951.9	1848.5	2800.4	2.534	6.269	**2.4**	
2.5	223.9	0.001197	0.0799	959.0	2601.2	962.0	1839.0	2800.9	2.554	6.254	**2.5**	
2.6	226.0	0.001201	0.0769	968.6	2601.5	971.7	1829.6	2801.4	2.574	6.239	**2.6**	
2.7	228.1	0.001205	0.0740	978.0	2601.8	981.2	1820.5	2801.7	2.592	6.224	**2.7**	
2.8	230.0	0.001209	0.0714	987.1	2602.1	990.5	1811.5	2802.0	2.611	6.210	**2.8**	
2.9	232.0	0.001213	0.0689	996.0	2602.3	999.5	1802.6	2802.2	2.628	6.197	**2.9**	
3.0	233.8	0.001216	0.0666	1004.7	2602.4	1008.4	1793.9	2802.3	2.646	6.184	**3.0**	**30** bar

Pressure in **MN/m²**

TABLE 8 (*cont.*). SATURATED WATER AND STEAM

PRESSURES FROM 3 MN/m² TO THE CRITICAL POINT (30 TO 221.2 BAR)

[0.1 MN/m² = 1 bar ≈ 14.5 lbf/in²]

Pressure MN/m²	Celsius temp., °C	Specific volume m³/kg		Specific internal energy kJ/kg		Specific enthalpy kJ/kg			Specific entropy kJ/kg K		Pressure MN/m²	
p	t	Water v_f	Steam v_g	Water u_f	Steam u_g	Water h_f	Evaporation h_{fg}	Steam h_g	Water s_f	Steam s_g	p	
3.0	233.8	0.001216	0.0666	1004.7	2602.4	1008.4	1793.9	2802.3	2.646	6.184	**3.0**	30 bar
3.2	237.4	0.001224	0.0624	1021.5	2602.5	1025.4	1776.9	2802.3	2.679	6.158	**3.2**	
3.4	240.9	0.001231	0.0587	1037.6	2602.5	1041.8	1760.3	2802.1	2.710	6.134	**3.4**	
3.6	244.2	0.001238	0.0554	1053.1	2602.2	1057.6	1744.2	2801.7	2.740	6.112	**3.6**	
3.8	247.3	0.001245	0.0524	1068.0	2601.9	1072.7	1728.4	2801.1	2.769	6.090	**3.8**	
4.0	250.3	0.001252	0.0497	1082.4	2601.3	1087.4	1712.9	2800.3	2.797	6.069	**4.0**	
4.2	253.2	0.001259	0.0473	1096.3	2600.7	1101.6	1697.8	2799.4	2.823	6.048	**4.2**	
4.4	256.0	0.001266	0.0451	1109.8	2599.9	1115.4	1682.9	2798.3	2.849	6.029	**4.4**	
4.6	258.8	0.001272	0.0430	1122.9	2599.1	1128.8	1668.3	2797.1	2.873	6.010	**4.6**	
4.8	261.4	0.001279	0.0412	1135.6	2598.1	1141.8	1653.9	2795.7	2.897	5.991	**4.8**	
5.0	263.9	0.001286	0.0394	1148.0	2597.0	1154.5	1639.7	2794.2	2.921	5.974	**5.0**	50 bar
5.2	266.4	0.001292	0.0378	1160.1	2595.9	1166.8	1625.7	2792.6	2.943	5.956	**5.2**	
5.4	268.8	0.001299	0.0363	1171.9	2594.6	1178.9	1611.9	2790.8	2.965	5.939	**5.4**	
5.6	271.1	0.001306	0.0349	1183.5	2593.3	1190.8	1598.2	2789.0	2.986	5.923	**5.6**	
5.8	273.3	0.001312	0.0337	1194.7	2591.9	1202.3	1584.7	2787.0	3.007	5.907	**5.8**	
6.0	275.6	0.001319	0.0324	1205.8	2590.4	1213.7	1571.3	2785.0	3.027	5.891	**6.0**	
6.2	277.7	0.001325	0.0313	1216.6	2588.8	1224.8	1558.0	2782.9	3.047	5.875	**6.2**	
6.4	279.8	0.001332	0.0302	1227.2	2587.2	1235.7	1544.9	2780.6	3.066	5.860	**6.4**	
6.6	281.8	0.001338	0.0292	1237.6	2585.5	1246.5	1531.9	2778.3	3.085	5.845	**6.6**	
6.8	283.8	0.001345	0.0283	1247.9	2583.7	1257.0	1518.9	2775.9	3.104	5.831	**6.8**	
7.0	285.8	0.001351	0.0274	1258.0	2581.8	1267.4	1506.0	2773.5	3.122	5.816	**7.0**	70 bar
7.2	287.7	0.001358	0.0265	1267.9	2579.9	1277.6	1493.3	2770.9	3.140	5.802	**7.2**	
7.4	289.6	0.001364	0.0257	1277.6	2578.0	1287.7	1480.5	2768.3	3.157	5.788	**7.4**	
7.6	291.4	0.001371	0.0249	1287.2	2575.9	1297.6	1467.9	2765.5	3.174	5.774	**7.6**	
7.8	293.2	0.001378	0.0242	1296.7	2573.8	1307.4	1455.3	2762.8	3.191	5.761	**7.8**	
8.0	295.0	0.001384	0.0235	1306.0	2571.7	1317.1	1442.8	2759.9	3.208	5.747	**8.0**	
8.2	296.7	0.001391	0.0229	1315.2	2569.5	1326.6	1430.3	2757.0	3.224	5.734	**8.2**	
8.4	298.4	0.001398	0.0222	1324.3	2567.2	1336.1	1417.9	2754.0	3.240	5.721	**8.4**	
8.6	300.1	0.001404	0.0216	1333.3	2564.9	1345.4	1405.5	2750.9	3.256	5.708	**8.6**	
8.8	301.7	0.001411	0.0210	1342.2	2562.6	1354.6	1393.2	2747.8	3.271	5.695	**8.8**	

Pressure in MN/m²

p	t	v_f	v_g	u_f	u_g	h_f	h_{fg}	h_g	s_f	s_g	p	
9.0	303.3	0.001418	0.02050	1351.0	2560.1	1363.7	1380.9	2744.6	3.287	5.682	**9.0**	
9.2	304.9	0.001425	0.01996	1359.7	2557.7	1372.8	1368.6	2741.4	3.302	5.669	**9.2**	
9.4	306.4	0.001432	0.01945	1368.2	2555.2	1381.7	1356.3	2738.0	3.317	5.657	**9.4**	
9.6	308.0	0.001439	0.01897	1376.7	2552.6	1390.6	1344.1	2734.7	3.332	5.644	**9.6**	
9.8	309.5	0.001446	0.01849	1385.2	2550.0	1399.3	1331.9	2731.2	3.346	5.632	**9.8**	
10.0	311.0	0.001453	0.01804	1393.5	2547.3	1408.0	1319.7	2727.7	3.361	5.620	**10.0**	**100** bar
10.5	314.6	0.001470	0.01698	1414.1	2540.4	1429.5	1289.2	2718.7	3.396	5.589	**10.5**	
11.0	318.0	0.001489	0.01601	1434.2	2533.2	1450.6	1258.7	2709.3	3.430	5.560	**11.0**	
11.5	321.4	0.001507	0.01511	1454.0	2525.7	1471.3	1228.2	2699.5	3.464	5.530	**11.5**	
12.0	324.6	0.001527	0.01428	1473.4	2517.8	1491.8	1197.4	2689.2	3.497	5.500	**12.0**	
12.5	327.8	0.001547	0.01351	1492.7	2509.4	1512.0	1166.4	2678.4	3.530	5.471	**12.5**	
13.0	330.8	0.001567	0.01280	1511.6	2500.6	1532.0	1135.0	2667.0	3.562	5.441	**13.0**	
13.5	333.8	0.001588	0.01213	1530.4	2491.3	1551.9	1103.1	2655.0	3.593	5.411	**13.5**	
14.0	336.6	0.001611	0.01150	1549.1	2481.4	1571.6	1070.7	2642.4	3.624	5.380	**14.0**	
14.5	339.4	0.001634	0.01090	1567.6	2471.0	1591.3	1037.7	2629.1	3.655	5.349	**14.5**	
15.0	342.1	0.001658	0.01034	1586.1	2459.9	1611.0	1004.0	2615.0	3.686	5.318	**15.0**	**150** bar
15.5	344.8	0.001683	0.00981	1604.6	2448.2	1630.7	969.6	2600.3	3.716	5.286	**15.5**	
16.0	347.3	0.001710	0.00931	1623.2	2436.0	1650.5	934.3	2584.9	3.747	5.253	**16.0**	
16.5	349.8	0.001739	0.00883	1641.8	2423.1	1670.5	898.3	2568.8	3.778	5.220	**16.5**	
17.0	352.3	0.001770	0.00837	1661.6	2409.3	1691.7	859.9	2551.6	3.811	5.185	**17.0**	
17.5	354.6	0.001803	0.00793	1681.8	2394.6	1713.3	820.0	2533.3	3.844	5.150	**17.5**	
18.0	357.0	0.001840	0.00750	1701.7	2378.9	1734.8	779.1	2513.9	3.877	5.113	**18.0**	
18.5	359.2	0.001881	0.00708	1721.7	2362.1	1756.5	736.6	2493.1	3.909	5.074	**18.5**	
19.0	361.4	0.001926	0.00668	1742.1	2343.8	1778.7	692.0	2470.6	3.943	5.033	**19.0**	
19.5	363.6	0.001977	0.00628	1763.2	2323.6	1801.8	644.2	2446.0	3.978	4.989	**19.5**	
20.0	365.7	0.00204	0.00588	1785.7	2300.8	1826.5	591.9	2418.4	4.015	4.941	**20.0**	**200** bar
20.5	367.8	0.00211	0.00546	1810.7	2274.4	1853.9	532.5	2386.4	4.056	4.887	**20.5**	
21.0	369.8	0.00220	0.00502	1840.0	2242.1	1886.3	461.3	2347.6	4.105	4.822	**21.0**	
21.5	371.8	0.00234	0.00451	1878.6	2198.1	1928.9	366.2	2295.2	4.169	4.737	**21.5**	
22.0	373.7	0.00267	0.00373	1952	2114	2011	185	2196	4.295	4.580	**22.0**	
22.12	374.15	0.00317	0.00317	2038	2038	2108	0	2108	4.444	4.444	22.12	**Critical point**

Pressure in **MN/m²**

TABLE 9. SPECIFIC ENTHALPY OF WATER AND STEAM

[0.1 MN/m² = 1 bar ≈ 14.5 lbf/in²]

Pressure/(MN/m²)	0	0.01	0.05	0.1	0.5	1	2	4	6	8	10	15	20	22.12 (Crit. isobar)	25	30	40	50	100	
Pressure/bar	0	0.1		1		10					100								1000	
Sat. Celsius temp., °C	—	45.8	81.3	99.6	151.8	179.9	212.4	250.3	275.6	295.0	311.0	342.1	365.7	374.15	—	—	—	—	—	
Sat. sp. enthalpy / kJ/kg, Water	—	191.8	340.6	417.5	640.1	762.6	908.6	1087.4	1213.7	1317.1	1408.0	1611.0	1826.5	2108	—	—	—	—	—	
Sat. sp. enthalpy / kJ/kg, Steam	—	2584.8	2646.0	2675.4	2747.5	2776.2	2797.2	2800.3	2785.0	2759.9	2727.7	2615.0	2418.4	2108	—	—	—	—	—	Celsius temp., °C
Celsius temp., °C	Specific enthalpy/(kJ/kg)																			
0	2502	0.0	0.0	0.1	0.5	1.0	2.0	4.0	6.1	8.1	10.1	15.1	20.1	22.2	25.1	30.0	39.7	49.3	95.9	0
25	2548	104.8	104.8	104.9	105.2	105.7	106.6	108.5	110.3	112.1	114.0	118.6	123.1	125.1	127.7	132.2	141.2	150.2	193.9	25
50	2595	2593	209.3	209.3	209.7	210.1	211.0	212.7	214.4	216.1	217.8	222.1	226.4	228.2	230.7	235.0	243.5	251.9	293.9	50
75	2642	2640	313.9	314.0	314.3	314.7	315.5	317.1	318.7	320.3	322.0	326.0	330.0	331.7	334.0	338.1	346.1	354.2	394.3	75
100	2689	2688	2683	2676	419.4	419.7	420.5	422.0	423.5	425.0	426.5	430.3	434.0	435.7	437.8	441.6	449.2	456.8	495.1	100
125	2736	2735	2731	2726	525.2	525.5	526.2	527.6	529.0	530.4	531.8	535.3	538.8	540.2	542.3	545.8	552.9	560.1	596.3	125
150	2784	2783	2780	2776	632.2	632.5	633.1	634.3	635.6	636.8	638.1	641.3	644.5	645.8	647.7	650.9	657.4	664.1	698.0	150
175	2832	2831	2829	2826	2800	741.1	741.7	742.7	743.8	744.9	746.0	748.7	751.5	752.7	754.4	757.2	763.1	769.1	800.4	175
200	2880	2880	2878	2875	2855	2827	852.6	853.4	854.2	855.1	855.9	858.1	860.4	861.4	862.8	865.2	870.2	875.4	903.5	200
225	2929	2928	2927	2925	2909	2886	2834	967.2	967.7	968.2	968.8	970.3	971.8	972.5	973.5	975.3	979.2	983.4	1007.7	225
250	2978	2977	2976	2975	2961	2943	2902	1085.8	1085.8	1085.8	1085.8	1086.2	1086.7	1087.0	1087.5	1088.4	1090.8	1093.6	1113.0	250
275	3027	3027	3026	3024	3013	2998	2965	2886	1210.8	1210.0	1209.2	1207.7	1206.6	1206.3	1205.9	1205.6	1205.7	1206.7	1219.9	275
300	3077	3077	3076	3074	3065	3052	3025	2962	2885	2787	1343.4	1338.3	1334.3	1332.8	1331.1	1328.7	1325.4	1323.7	1328.7	300
325	3127	3127	3126	3125	3116	3106	3083	3031	2970	2899	2811	1486.0	1475.5	1471.8	1467.4	1461.1	1452.0	1446.0	1439.1	325
350	3177	3177	3177	3176	3168	3159	3139	3095	3046	2990	2926	2695	1647.1	1636.5	1625.0	1609.9	1589.6	1576.3	1550.5	350
375	3228	3228	3228	3227	3220	3211	3194	3156	3115	3069	3019	2862	2604	2319	1849	1791	1742	1716	1671	375
400	3280	3280	3279	3278	3272	3264	3249	3216	3180	3142	3100	2979	2820	2733	2582	2162	1934	1878	1798	400
425	3331	3331	3331	3330	3325	3317	3303	3274	3243	3209	3174	3075	2957	2899	2810	2619	2208	2068	1924	425
450	3384	3384	3383	3382	3377	3371	3358	3331	3303	3274	3244	3160	3064	3020	2954	2826	2516	2293	2051	450
475	3436	3436	3436	3435	3430	3424	3412	3388	3363	3337	3310	3237	3157	3120	3068	2969	2743	2522	2181	475
500	3489	3489	3489	3488	3484	3478	3467	3445	3422	3399	3375	3311	3241	3210	3166	3085	2907	2723	2316	500
550	3597	3596	3596	3596	3592	3587	3578	3559	3539	3520	3500	3448	3394	3370	3337	3277	3152	3021	2594	550
600	3706	3706	3705	3705	3702	3697	3689	3673	3656	3640	3623	3580	3536	3516	3490	3443	3346	3248	2857	600
650	3816	3816	3816	3816	3813	3809	3802	3788	3774	3759	3745	3708	3671	3655	3633	3595	3517	3439	3105	650
700	3929	3929	3929	3928	3926	3923	3916	3904	3892	3879	3867	3835	3804	3790	3772	3740	3675	3610	3324	700
750	4043	4043	4043	4042	4040	4038	4032	4021	4011	4000	3989	3962	3935	3923	3908	3880	3825	3771	3526	750
800	4159	4159	4159	4158	4156	4154	4149	4140	4131	4121	4112	4089	4065	4055	4042	4018	3972	3925	3714	800

TABLE 10. SPECIFIC ENTROPY OF WATER AND STEAM

[0.1 MN/m² = 1 bar ≈ 14.5 lbf/in²]

Pressure/(MN/m²)	0	0.01	0.05	0.1	0.5	1	2	4	6	8	10	15	20	Crit. isobar 22.12	25	30	40	50	100	
Pressure/bar	0	0.1		1		10					100								1000	
Sat. Celsius temp., °C	—	45.8	81.3	99.6	151.8	179.9	212.4	250.3	275.6	295.0	311.0	342.1	365.7	374.15	—	—	—	—	—	
Sat. sp. entropy / kJ/kg K — Water	—	0.649	1.091	1.303	1.860	2.138	2.447	2.797	3.027	3.208	3.361	3.686	4.015	4.444	—	—	—	—	—	
Sat. sp. entropy / kJ/kg K — Steam	—	8.151	7.595	7.360	6.819	6.583	6.337	6.069	5.891	5.747	5.620	5.318	4.941	4.444	—	—	—	—	—	
Celsius temp., °C	Specific entropy/(kJ/kg K)																			Celsius temp., °C
0		0.000	0.000	0.000	0.000	0.000	0.000	0.000	0.000	0.000	0.001	0.001	0.001	0.001	0.001	0.001	0.000	0.000	−0.007	0
25		0.367	0.367	0.367	0.367	0.367	0.366	0.366	0.365	0.365	0.364	0.363	0.362	0.361	0.360	0.359	0.356	0.353	0.336	25
50		8.176	0.703	0.703	0.703	0.703	0.703	0.702	0.701	0.700	0.699	0.697	0.694	0.693	0.692	0.690	0.685	0.681	0.658	50
75		8.317	1.015	1.015	1.015	1.015	1.014	1.013	1.012	1.010	1.009	1.006	1.003	1.002	1.000	0.997	0.991	0.985	0.958	75
100		8.449	7.695	7.362	1.307	1.306	1.305	1.304	1.302	1.301	1.299	1.295	1.292	1.290	1.288	1.284	1.277	1.270	1.237	100
125		8.572	7.822	7.492	1.581	1.581	1.580	1.578	1.576	1.574	1.572	1.568	1.563	1.561	1.559	1.555	1.546	1.538	1.500	125
150		8.689	7.941	7.614	1.842	1.841	1.840	1.838	1.836	1.833	1.831	1.826	1.821	1.819	1.816	1.811	1.801	1.791	1.748	150
175		8.799	8.053	7.728	6.940	2.091	2.089	2.087	2.084	2.081	2.079	2.073	2.066	2.064	2.061	2.055	2.043	2.032	1.983	175
200		8.905	8.159	7.835	7.059	6.692	2.330	2.327	2.324	2.321	2.318	2.310	2.303	2.300	2.296	2.289	2.276	2.263	2.207	200
225		9.005	8.260	7.937	7.169	6.815	6.412	2.561	2.557	2.554	2.550	2.541	2.532	2.529	2.524	2.516	2.500	2.486	2.421	225
250		9.101	8.356	8.034	7.272	6.926	6.545	2.793	2.789	2.784	2.779	2.768	2.757	2.753	2.747	2.737	2.719	2.701	2.627	250
275		9.193	8.449	8.127	7.369	7.029	6.663	6.229	3.022	3.016	3.010	2.995	2.981	2.976	2.968	2.956	2.933	2.913	2.827	275
300		9.282	8.538	8.217	7.461	7.125	6.770	6.364	6.069	5.794	3.249	3.228	3.209	3.201	3.192	3.176	3.147	3.121	3.021	300
325	(Infinite)	9.368	8.624	8.303	7.550	7.216	6.868	6.482	6.215	5.986	5.761	3.480	3.450	3.439	3.424	3.402	3.363	3.330	3.210	325
350		9.450	8.707	8.386	7.634	7.303	6.960	6.587	6.339	6.135	5.949	5.447	3.731	3.708	3.683	3.646	3.589	3.544	3.392	350
375		9.531	8.787	8.466	7.716	7.386	7.047	6.683	6.448	6.260	6.095	5.710	5.230	4.769	4.034	3.930	3.828	3.764	3.582	375
400		9.608	8.865	8.544	7.795	7.467	7.130	6.773	6.546	6.369	6.218	5.888	5.559	5.401	5.145	4.490	4.119	4.008	3.774	400
425		9.684	8.940	8.620	7.871	7.544	7.209	6.858	6.637	6.468	6.326	6.028	5.758	5.643	5.478	5.158	4.519	4.285	3.959	425
450		9.757	9.014	8.693	7.945	7.619	7.286	6.939	6.723	6.560	6.424	6.147	5.909	5.813	5.682	5.449	4.951	4.603	4.137	450
475		9.829	9.085	8.765	8.018	7.692	7.360	7.016	6.804	6.645	6.515	6.252	6.035	5.949	5.837	5.645	5.261	4.914	4.314	475
500		9.898	9.155	8.835	8.088	7.763	7.432	7.091	6.882	6.726	6.600	6.349	6.146	6.067	5.965	5.797	5.476	5.178	4.491	500
550		10.033	9.290	8.970	8.223	7.899	7.571	7.233	7.029	6.878	6.756	6.521	6.337	6.268	6.180	6.039	5.784	5.552	4.839	550
600		10.162	9.419	9.098	8.353	8.029	7.702	7.368	7.166	7.019	6.901	6.676	6.504	6.441	6.360	6.234	6.013	5.821	5.151	600
650		10.285	9.542	9.222	8.477	8.154	7.828	7.496	7.297	7.152	7.037	6.819	6.655	6.595	6.520	6.403	6.204	6.033	5.427	650
700		10.404	9.661	9.341	8.596	8.273	7.949	7.619	7.422	7.279	7.166	6.954	6.795	6.738	6.666	6.556	6.370	6.214	5.658	700
750		10.518	9.775	9.455	8.711	8.389	8.065	7.736	7.541	7.400	7.289	7.081	6.927	6.871	6.803	6.697	6.521	6.375	5.860	750
800		10.628	9.886	9.565	8.821	8.500	8.176	7.849	7.655	7.516	7.406	7.201	7.051	6.997	6.931	6.829	6.661	6.522	6.040	800

TABLE 11. DENSITY OF WATER AND STEAM

[0.1 MN/m² = 1 bar ≈ 14.5 lbf/in²]

Pressure/(MN/m²)	0	0.01	0.05	0.1	0.5	1	2	4	6	8	10	15	20	Crit. isobar 22.12	25	30	40	50	100	
Pressure/bar	0	0.1		1		10					100								1000	
Sat. Celsius temp., °C	—	45.8	81.3	99.6	151.8	179.9	212.4	250.3	275.6	295.0	311.0	342.1	365.7	374.15	—	—	—	—	—	
Sat. density kg/m³ Water	—	990	971	958	915	887	850	799	758	722	688	603	491	315	—	—	—	—	—	
Sat. density kg/m³ Steam	—	0.0681	0.309	0.590	2.67	5.15	10.05	20.10	30.8	42.5	55.4	96.7	170.2	315	—	—	—	—	—	Celsius temp., °C
Celsius temp., °C		Density/(kg/m³)																		
0		1000	1000	1000	1000	1000	1001	1002	1003	1004	1005	1007	1010	1011	1012	1014	1019	1024	1046	0
25		997	997	997	997	998	998	999	1000	1001	1002	1004	1006	1007	1008	1010	1014	1018	1038	25
50		0.0672	988	988	988	988	989	990	991	992	992	994	997	997	999	1001	1005	1009	1027	50
75		0.0624	975	975	975	975	976	976	977	978	979	981	984	984	986	988	992	996	1015	75
100		0.0582	0.293	0.590	958	959	959	960	961	962	963	965	967	968	970	972	976	980	1000	100
125		0.0545	0.274	0.550	939	939	940	941	942	943	944	946	949	950	951	954	958	963	984	125
150		0.0512	0.257	0.516	917	917	918	919	920	921	922	925	928	929	930	933	938	943	965	150
175		0.0484	0.243	0.487	2.504	892	893	894	896	897	898	901	904	906	907	910	916	921	946	175
200		0.0458	0.230	0.460	2.353	4.86	865	867	868	870	871	875	878	880	882	885	891	897	924	200
225		0.0435	0.218	0.437	2.223	4.55	9.64	835	837	839	841	845	849	851	853	857	864	871	901	225
250		0.0414	0.207	0.416	2.108	4.30	8.97	799	802	804	806	811	817	819	821	826	835	843	877	250
275		0.0395	0.198	0.396	2.006	4.07	8.43	18.33	759	762	765	773	779	782	785	791	802	811	850	275
300		0.0378	0.189	0.379	1.914	3.88	7.97	17.00	27.67	41.2	715	726	735	739	743	751	765	777	823	300
325	(Zero)	0.0362	0.181	0.363	1.830	3.70	7.57	15.93	25.41	36.5	50.4	665	680	685	692	703	722	738	793	325
350		0.0348	0.174	0.348	1.754	3.54	7.22	15.05	23.68	33.4	44.6	87.2	600	612	625	644	671	693	761	350
375		0.0334	0.167	0.335	1.685	3.40	6.90	14.29	22.28	31.0	40.8	72.0	130.5	218.4	504	558	609	641	727	375
400		0.0322	0.161	0.322	1.620	3.26	6.62	13.63	21.11	29.1	37.9	63.9	100.5	122.6	166.3	353.3	523.8	578.3	691.4	400
425		0.0310	0.155	0.311	1.561	3.14	6.36	13.04	20.09	27.6	35.6	58.4	87.2	102.2	126.8	188.3	392.4	498.3	654.1	425
450		0.0300	0.150	0.300	1.506	3.03	6.12	12.51	19.19	26.2	33.6	54.2	78.7	90.7	109.0	148.5	272.1	401.3	613.9	450
475		0.0290	0.145	0.290	1.455	2.92	5.90	12.02	18.39	25.0	32.0	50.9	72.5	82.8	97.9	128.3	210.3	316.0	571.4	475
500		0.0280	0.140	0.280	1.407	2.83	5.70	11.58	17.67	24.0	30.5	48.1	67.7	76.8	89.9	115.2	178.1	257.6	528.2	500
550		0.0263	0.132	0.263	1.320	2.65	5.33	10.80	16.41	22.2	28.1	43.7	60.4	68.0	78.6	98.4	143.2	195.6	445.3	550
600		0.0248	0.124	0.248	1.244	2.49	5.01	10.13	15.34	20.7	26.1	40.2	55.1	61.6	70.8	87.4	123.6	163.6	374.8	600
650		0.0235	0.117	0.235	1.176	2.36	4.73	9.54	14.42	19.4	24.4	37.4	50.8	56.7	64.9	79.5	110.5	143.7	322.0	650
700		0.0223	0.111	0.223	1.115	2.23	4.48	9.02	13.61	18.3	23.0	35.0	47.4	52.7	60.1	73.3	100.7	129.5	282.8	700
750		0.0212	0.106	0.212	1.060	2.12	4.26	8.55	12.89	17.3	21.7	32.9	44.4	49.4	56.2	68.2	93.0	118.8	253.0	750
800		0.0202	0.101	0.202	1.011	2.02	4.05	8.14	12.26	16.4	20.6	31.2	41.9	46.6	52.9	64.0	86.8	110.2	230.4	800

Note: **Density** is tabulated here, instead of specific volume, since interpolation between pressures is thereby facilitated.

TABLE 12. SPECIFIC INTERNAL ENERGY OF WATER AND STEAM

[0.1 MN/m² = 1 bar ≈ 14.5 lbf/in²]

Pressure/(MN/m²)	0	0.01	0.05	0.1	0.5	1	2	4	6	8	10	15	20	Crit. isobar 22.12	25	30	40	50	100	
Pressure/bar	0	0.1		1		10					100								1000	
Sat. Celsius temp., °C	—	45.8	81.3	99.6	151.8	179.9	212.4	250.3	275.6	295.0	311.0	342.1	365.7	374.15	—	—	—	—	—	
Sat. sp. int. energy kJ/kg, Water	—	191.8	340.5	417.4	639.6	761.5	906.2	1082.4	1205.8	1306.0	1393.5	1586.1	1785.7	2037.8	—	—	—	—	—	
Sat. sp. int. energy kJ/kg, Steam	—	2438.0	2484.0	2506.1	2560.2	2581.9	2598.2	2601.3	2590.4	2571.7	2547.3	2459.9	2300.8	2037.8	—	—	—	—	—	
Celsius temp., °C	Specific internal energy/(kJ/kg)																			Celsius temp., °C
0	2376	0.0	0.0	0.0	0.0	0.0	0.0	0.0	0.1	0.1	0.2	0.3	0.3	0.4	0.4	0.4	0.5	0.5	0.2	0
25	2411	104.8	104.8	104.8	104.7	104.7	104.6	104.5	104.3	104.2	104.0	103.6	103.3	103.1	102.9	102.5	101.8	101.1	97.5	25
50	2446	2444	209.2	209.2	209.2	209.1	209.0	208.6	208.3	208.1	207.8	207.0	206.3	206.0	205.7	205.0	203.7	202.4	196.5	50
75	2481	2480	313.9	313.9	313.8	313.7	313.5	313.0	312.6	312.2	311.7	310.7	309.7	309.2	308.7	307.7	305.8	304.0	295.7	75
100	2517	2516	2512	2507	418.8	418.7	418.4	417.8	417.3	416.7	416.1	414.7	413.4	412.8	412.1	410.8	408.2	405.8	395.1	100
125	2552	2552	2549	2545	524.6	524.5	524.1	523.3	522.6	521.9	521.2	519.4	517.7	517.0	516.0	514.3	511.2	508.1	494.6	125
150	2589	2588	2586	2583	631.6	631.4	630.9	630.0	629.1	628.2	627.3	625.0	622.9	622.0	620.8	618.7	614.8	611.0	594.4	150
175	2625	2625	2623	2620	2601	740.0	739.4	738.2	737.1	736.0	734.8	732.1	729.4	728.3	726.8	724.3	719.4	714.8	694.6	175
200	2662	2661	2660	2658	2643	2621	850.2	848.8	847.3	845.9	844.4	841.0	837.7	836.3	834.4	831.3	825.3	819.7	795.3	200
225	2699	2699	2697	2696	2684	2667	2627	962.4	960.6	958.7	956.9	952.5	948.3	946.5	944.2	940.3	932.9	926.0	896.7	225
250	2736	2736	2735	2734	2724	2710	2679	1080.8	1078.3	1075.8	1073.4	1067.7	1062.2	1060.0	1057.0	1052.1	1042.8	1034.3	999.0	250
275	2774	2774	2773	2772	2764	2753	2728	2668	1202.9	1199.5	1196.1	1188.3	1181.0	1178.0	1174.1	1167.7	1155.8	1145.1	1102.4	275
300	2812	2812	2811	2811	2803	2794	2774	2727	2668	2593	1329.4	1317.6	1307.1	1302.9	1297.5	1288.7	1273.1	1259.3	1207.1	300
325	2851	2851	2850	2849	2843	2835	2818	2780	2734	2680	2612	1463.5	1446.0	1439.5	1431.3	1418.5	1396.6	1378.2	1313.0	325
350	2890	2890	2889	2889	2883	2876	2862	2829	2792	2750	2702	2523	1613.7	1600.3	1585.0	1563.3	1530.0	1504.1	1419.0	350
375	2929	2929	2929	2928	2923	2917	2904	2876	2846	2811	2773	2653	2450	2217	1799	1737	1676	1638	1534	375
400	2969	2969	2969	2968	2964	2958	2946	2922	2896	2867	2836	2744	2622	2553	2432	2077	1858	1791	1653	400
425	3009	3009	3009	3008	3004	2999	2989	2967	2944	2919	2893	2818	2728	2683	2613	2460	2107	1967	1771	425
450	3050	3050	3049	3049	3045	3041	3031	3011	2991	2969	2946	2883	2810	2776	2725	2623	2369	2169	1888	450
475	3091	3091	3091	3090	3087	3082	3073	3056	3037	3018	2997	2943	2881	2853	2813	2735	2553	2364	2006	475
500	3132	3132	3132	3132	3128	3124	3116	3100	3083	3065	3047	2999	2946	2922	2888	2825	2682	2529	2127	500
550	3217	3217	3216	3216	3213	3210	3202	3188	3174	3159	3144	3105	3063	3045	3019	2972	2872	2765	2369	550
600	3303	3303	3302	3302	3300	3296	3290	3278	3265	3252	3240	3207	3172	3157	3137	3100	3023	2943	2591	600
650	3390	3390	3390	3390	3388	3385	3379	3368	3357	3346	3335	3307	3278	3265	3248	3218	3155	3091	2795	650
700	3480	3480	3480	3479	3477	3475	3470	3460	3451	3441	3431	3407	3382	3371	3356	3330	3278	3224	2971	700
750	3571	3571	3571	3570	3569	3567	3562	3554	3545	3537	3528	3507	3485	3476	3463	3441	3396	3350	3131	750
800	3663	3663	3663	3663	3662	3660	3656	3649	3641	3634	3626	3607	3588	3580	3569	3550	3511	3471	3280	800

REFRIGERANT TABLES

In Tables 13–16, the arbitrary datum state for zero enthalpy and entropy is that of the saturated liquid at a Celsius temperature of −40 °C

TABLE 13. PROPERTIES OF REFRIGERANT-12, CCl_2F_2

[0.1 MN/m² = 1 bar ≈ 14.5 lbf/in²]

Saturation Celsius temp., °C	Pressure MN/m²	Saturated: Specific volume m³/kg		Saturated: Specific enthalpy kJ/kg		Saturated: Specific entropy kJ/kg K		Superheated: By 20 K		Superheated: By 40 K	
		Liquid	Vapour	Liquid	Vapour	Liquid	Vapour				
t	p	v_f	v_g	h_f	h_g	s_f	s_g	h	s	h	s
−40	0.0641	0.00066	0.2421	zero	169.6	zero	0.7274	180.8	0.7737	192.4	0.8178
−35	0.0806	0.00067	0.1955	4.4	171.9	0.0187	0.7220	183.3	0.7681	195.1	0.8120
−30	0.1003	0.00067	0.1595	8.9	174.2	0.0371	0.7171	185.8	0.7631	197.8	0.8068
−25	0.1236	0.00068	0.1313	13.3	176.5	0.0552	0.7127	188.3	0.7586	200.4	0.8021
−20	0.1508	0.00069	0.1089	17.8	178.7	0.0731	0.7088	190.8	0.7546	203.1	0.7979
−15	0.1825	0.00069	0.0911	22.3	181.0	0.0906	0.7052	193.2	0.7510	205.7	0.7942
−10	0.219	0.00070	0.0767	26.9	183.2	0.1080	0.7020	195.7	0.7477	208.3	0.7909
−5	0.261	0.00071	0.0650	31.4	185.4	0.1251	0.6991	198.1	0.7449	210.9	0.7879
0	0.308	0.00072	0.0554	36.1	187.5	0.1420	0.6966	200.5	0.7423	213.5	0.7853
5	0.362	0.00072	0.0475	40.7	189.7	0.1587	0.6942	202.9	0.7401	216.1	0.7830
10	0.423	0.00073	0.0409	45.4	191.7	0.1752	0.6921	205.2	0.7381	218.6	0.7810
15	0.491	0.00074	0.0354	50.1	193.8	0.1915	0.6902	207.5	0.7363	221.2	0.7792
20	0.567	0.00075	0.0308	54.9	195.8	0.2078	0.6885	209.8	0.7348	223.7	0.7777
25	0.651	0.00076	0.0269	59.7	197.7	0.2239	0.6869	212.1	0.7334	226.1	0.7763
30	0.745	0.00077	0.0235	64.6	199.6	0.2399	0.6854	214.3	0.7321	228.6	0.7751
35	0.847	0.00079	0.0206	69.5	201.5	0.2559	0.6839	216.4	0.7310	231.0	0.7741
40	0.960	0.00080	0.0182	74.6	203.2	0.2718	0.6825	218.5	0.7300	233.4	0.7732
45	1.084	0.00081	0.0160	79.7	204.9	0.2877	0.6812	220.6	0.7291	235.7	0.7724
50	1.219	0.00083	0.0142	84.9	206.5	0.3037	0.6797	222.6	0.7282	238.0	0.7718

Freezing point at 1 atm = −155.0 °C
Critical point: Celsius temp. = 112.0 °C
pressure = 4.115 MN/m² (41.15 bar)

TABLE 14. PROPERTIES OF METHYL CHLORIDE, CH_3Cl

[0.1 MN/m² = 1 bar ≈ 14.5 lbf/in²]

Saturation Celsius temp., °C	Pressure MN/m²	Saturated: Specific volume m³/kg		Saturated: Specific enthalpy kJ/kg		Saturated: Specific entropy kJ/kg K		Superheated: By 30 K		Superheated: By 60 K	
		Liquid	Vapour	Liquid	Vapour	Liquid	Vapour				
t	p	v_f	v_g	h_f	h_g	s_f	s_g	h	s	h	s
−40	0.0474	0.00097	0.794	**zero**	443.5	**zero**	1.902	465.9	1.993	489.6	2.078
−35	0.0607	0.00098	0.632	7.5	446.5	0.032	1.875	469.2	1.965	493.2	2.050
−30	0.0767	0.00099	0.508	14.9	449.4	0.063	1.850	472.6	1.940	496.9	2.023
−25	0.0960	0.00100	0.412	22.5	452.3	0.094	1.826	475.9	1.916	500.5	2.000
−20	0.119	0.00100	0.338	30.1	455.2	0.124	1.803	479.1	1.893	504.1	1.977
−15	0.146	0.00101	0.279	37.7	457.9	0.154	1.782	482.4	1.871	507.7	1.955
−10	0.177	0.00102	0.233	45.4	460.7	0.183	1.762	485.5	1.851	511.3	1.935
−5	0.214	0.00103	0.195	53.1	463.3	0.212	1.742	488.6	1.832	514.8	1.915
0	0.256	0.00104	0.165	60.9	465.8	0.241	1.724	491.6	1.813	518.2	1.897
5	0.304	0.00105	0.140	68.7	468.2	0.269	1.706	494.6	1.796	521.7	1.880
10	0.358	0.00106	0.120	76.6	470.5	0.297	1.689	497.4	1.779	525.1	1.864
15	0.420	0.00107	0.103	84.5	472.7	0.325	1.673	500.3	1.764	528.4	1.848
20	0.490	0.00109	0.089	92.5	474.9	0.353	1.657	503.0	1.749	531.7	1.834
25	0.567	0.00110	0.077	100.5	476.8	0.379	1.642	505.7	1.735	535.0	1.820
30	0.653	0.00111	0.068	108.6	478.7	0.406	1.627	508.4	1.721	538.2	1.807
35	0.748	0.00112	0.059	116.7	480.4	0.433	1.613	511.0	1.708	541.3	1.794
40	0.852	0.00113	0.052	124.8	482.1	0.459	1.600	513.5	1.696	544.5	1.783
45	0.967	0.00115	0.046	133.0	483.6	0.485	1.587	515.9	1.684	547.6	1.772
50	1.092	0.00116	0.041	141.3	485.0	0.511	1.575	518.2	1.674	550.6	1.761

Freezing point at 1 atm = −97.6 °C
Critical point: Celsius temp. = 143.1 °C
pressure = 6.68 MN/m² (66.8 bar)

TABLE 15. PROPERTIES OF AMMONIA, NH_3

[0.1 MN/m² = 1 bar ≈ 14.5 lbf/in²]

Saturation Celsius temp., °C	Pressure MN/m²	Saturated: Specific volume m³/kg		Saturated: Specific enthalpy kJ/kg		Saturated: Specific entropy kJ/kg K		Superheated: By 50 K		Superheated: By 100 K	
		Liquid	Vapour	Liquid	Vapour	Liquid	Vapour				
t	p	v_f	v_g	h_f	h_g	s_f	s_g	h	s	h	s
−40	0.0718	0.00145	1.552	zero	1390	zero	5.963	1499	6.387	1606	6.736
−35	0.0932	0.00146	1.216	22.3	1398	0.095	5.872	1508	6.292	1616	6.639
−30	0.1196	0.00148	0.963	44.7	1406	0.188	5.785	1517	6.203	1626	6.547
−25	0.1516	0.00149	0.772	67.2	1413	0.279	5.703	1526	6.119	1636	6.461
−20	0.190	0.00150	0.624	89.8	1420	0.368	5.624	1535	6.039	1646	6.379
−15	0.236	0.00152	0.509	112.3	1426	0.457	5.549	1543	5.963	1656	6.301
−10	0.291	0.00153	0.418	135.4	1433	0.544	5.477	1552	5.891	1665	6.227
−5	0.355	0.00155	0.347	158.2	1439	0.630	5.407	1560	5.822	1675	6.157
0	0.429	0.00157	0.289	181.2	1444	0.715	5.340	1568	5.756	1684	6.090
5	0.516	0.00158	0.243	204.5	1450	0.799	5.276	1576	5.694	1693	6.027
10	0.615	0.00160	0.206	227.7	1454	0.881	5.214	1583	5.634	1702	5.967
15	0.728	0.00162	0.175	251.4	1459	0.963	5.154	1590	5.576	1711	5.909
20	0.857	0.00164	0.149	275.2	1463	1.044	5.095	1597	5.521	1719	5.853
25	1.001	0.00166	0.128	298.9	1466	1.124	5.039	1604	5.468	1728	5.800
30	1.167	0.00168	0.111	323.1	1469	1.204	4.984	1610	5.417	1736	5.750
35	1.350	0.00170	0.096	347.5	1471	1.282	4.930	1616	5.368	1744	5.702
40	1.554	0.00173	0.083	371.5	1473	1.360	4.877	1622	5.321	1752	5.655
45	1.782	0.00175	0.073	396.8	1474	1.437	4.825	1628	5.275	1760	5.610
50	2.033	0.00178	0.063	421.9	1475	1.515	4.773	1633	5.230	1767	5.567

Freezing point at 1 atm = −77.7 °C
Critical point: Celsius temp. = 132.4 °C
pressure = 11.30 MN/m² (113.0 bar)

TABLE 16. PROPERTIES OF CARBON DIOXIDE, CO_2

[0.1 MN/m² = 1 bar ≈ 14.5 lbf/in²]

Saturation Celsius temp., °C	Pressure MN/m²	Saturated: Specific volume m³/kg		Saturated: Specific enthalpy kJ/kg		Saturated: Specific entropy kJ/kg K		Superheated: By 30 K		Superheated: By 60 K		
		Liquid	Vapour	Liquid	Vapour	Liquid	Vapour					
t	p	v_f	v_g	h_f	h_g	s_f	s_g	h	s	h	s	
−40	1.005	0.00090	0.0382	**zero**	321.1	**zero**	1.377	355.4	1.507	383.0	1.611	
−35	1.20	0.00091	0.0320	9.7	322.2	0.039	1.352	356.9	1.485	385.6	1.588	
−30	1.43	0.00093	0.0270	19.5	323.1	0.079	1.328	358.7	1.464	388.0	1.566	
−25	1.68	0.00095	0.0229	29.5	323.7	0.119	1.304	360.4	1.442	390.3	1.545	
−20	1.97	0.00097	0.0195	39.7	323.7	0.158	1.280	361.8	1.421	392.5	1.525	
−15	2.29	0.00099	0.0166	50.2	323.2	0.198	1.256	363.0	1.401	394.5	1.505	
−10	2.65	0.00102	0.0142	60.9	322.3	0.238	1.231	363.9	1.381	396.2	1.486	
−5	3.04	0.00105	0.0122	72.0	320.5	0.278	1.205	364.6	1.361	397.8	1.467	
0	3.48	0.00108	0.0104	83.7	318.1	0.320	1.178	364.9	1.342	399.3	1.449	
5	3.97	0.00111	0.00879	96.0	312.9	0.364	1.143	364.9	1.322	400.4	1.431	
10	4.50	0.00116	0.00743	109.1	307.2	0.407	1.107	364.7	1.302	401.4	1.414	
15	5.08	0.00121	0.00623	123.3	301.0	0.454	1.071	364.0	1.282	402.2	1.396	
20	5.73	0.00129	0.00516	139.1	292.3	0.506	1.028	362.9	1.261	402.7	1.379	
25	6.44	0.00140	0.00413	159.7	279.9	0.573	0.976	361.5	1.241	403.0	1.362	
30	7.21	0.00169	0.00294	191.2	253.1	0.682	0.886	359.6	1.220	402.9	1.345	
31.05	7.38	0.00214	0.00214	223.0	223.0	0.780	0.780	359.1	1.216	402.9	1.341	**Critical point**

Triple point: temperature = −56.6 °C, pressure = 0.518 MN/m²
Sublimation point at 1 atm = −78.5 °C

AIR AT LOW TEMPERATURES

In Tables 17–19, the arbitrary datum state for zero enthalpy and entropy is that of the saturated liquid at a pressure of **0.1** MN/m^2 (1 bar)

TABLE 17. SATURATED LIQUID AND VAPOUR (AIR)

Warning: This table lists *absolute* temperatures.

[0.1 MN/m^2 = 1 bar ≈ 14.5 lbf/in^2]

Pressure MN/m^2	Temperature K		Specific volume m^3/kg		Specific enthalpy kJ/kg		Specific entropy kJ/kg K		Pressure MN/m^2	
p	Liquid (bubble)	Vapour (dew)	Liquid v_f	Vapour v_g	Liquid h_f	Vapour h_g	Liquid s_f	Vapour s_g	p	
0.1	78.7	81.7	0.00114	0.224	**zero**	205.3	**zero**	2.559	**0.1**	**1** bar
0.2	85.4	88.1	0.00118	0.119	11.9	210.5	0.143	2.433	**0.2**	
0.3	90.7	92.5	0.00122	0.081	20.0	213.3	0.234	2.354	**0.3**	
0.5	96.2	98.5	0.00127	0.050	31.7	215.9	0.357	2.249	**0.5**	
0.7	100.8	103.0	0.00132	0.036	41.0	217.0	0.448	2.175	**0.7**	
1.0	106.2	108.1	0.00138	0.0251	53.1	217.1	0.561	2.092	**1.0**	**10** bar
1.5	113.1	114.7	0.00149	0.0163	70.7	215.5	0.715	1.987	**1.5**	
2.0	118.5	119.8	0.00160	0.0116	86.6	211.7	0.847	1.898	**2.0**	
2.5	123.0	124.1	0.00173	0.0087	101.6	206.3	0.964	1.812	**2.5**	
3.0	127.0	127.8	0.00190	0.0066	116.8	199.0	1.077	1.723	**3.0**	
3.5	130.1	131.1	0.00221	0.0048	134.2	187.7	1.212	1.621	**3.5**	
3.766	132.5 max. point of contact		0.00313		164.4		1.430		3.766	**Critical point**
3.774 (max.)	132.4 plait point		0.00305		162.7		1.418		3.774 (max.)	**Critical point**

TABLE 18. SPECIFIC ENTHALPY OF LIQUID AND VAPOUR (AIR)

Warning: This table lists *absolute* temperatures.

[0.1 MN/m² = 1 bar ≈ 14.5 lbf/in²]

Pressure MN/m²	Sat. temp. K Liq.	Sat. temp. K Vap.	Temperature/K ... 90	100	110	120	125	130	135	140	145	150	160	170	180	190	200	250	300	Pressure MN/m²	
			Specific enthalpy kJ/kg																		
0.1	78.7	81.7	214.0	224.4	234.8	245.2	250.4	255.6	260.7	265.9	271.1	276.2	286.5	296.7	306.9	317.1	327.2	377.7	427.8	0.1	1 bar
0.5	96.2	98.5		217.9	229.4	240.7	246.2	251.7	257.1	262.4	267.8	273.1	283.7	294.2	304.7	315.1	325.4	376.5	427.2	0.5	
1	106.2	108.1			219.8	233.1	239.4	245.5	251.4	257.2	263.0	268.7	279.9	290.9	301.7	312.4	323.1	375.0	426.2	1	10 bar
2	118.5	119.8					221.4	230.0	237.8	245.2	252.2	258.9	271.7	283.8	295.6	307.0	318.1	372.0	424.2	2	
3	127.0	127.8				81.0	103.5	206.6	219.7	230.4	239.5	247.8	262.8	276.4	289.2	301.4	313.1	368.9	422.2	3	
4						75.6	94.0	117.7	166.5	209.1	223.6	234.6	252.7	268.3	282.4	295.6	308.1	365.8	420.2	4	
5						73.1	89.7	109.2	134.4	165.4	193.5	215.4	241.1	259.6	275.4	289.7	302.9	362.7	418.1	5	
10						70.1	83.9	98.1	112.8	128.0	143.7	159.5	190.3	217.2	239.8	259.3	276.7	347.6	408.5	10	100 bar
15						70.4	83.1	95.8	108.6	121.5	134.5	147.5	172.8	196.5	218.5	238.8	257.5	335.1	400.5	15	
20						72.1	84.0	95.8	107.6	119.4	131.1	142.7	165.5	187.3	208.0	227.6	246.0	325.9	394.3	20	
25						75.5	86.9	98.2	109.5	120.6	131.5	142.3	163.4	183.8	203.3	221.9	239.8	319.7	389.8	25	
30						80.0	91.2	102.2	113.0	123.6	134.1	144.3	164.1	183.2	201.7	219.5	236.8	315.9	386.8	30	
35													166.8	184.8	202.4	219.5	236.2	314.1	385.2	35	
40													170.5	187.7	204.5	220.9	237.1	313.7	384.9	40	
45													175.0	191.4	207.6	223.5	239.1	314.5	385.4	45	
50													179.7	195.4	211.0	226.5	241.7	315.7	386.3	50	500 bar
60															218.3	233.2	247.9	319.9	389.3	60	
70																240.4	254.8	325.2	393.6	70	
80																248.3	262.3	331.5	399.1	80	
90																		338.4	405.4	90	
100																		345.5	412.1	100	1000 bar

TABLE 19. SPECIFIC ENTROPY OF LIQUID AND VAPOUR (AIR)

Warning: This table lists *absolute* temperatures.

[0.1 MN/m² = 1 bar ≈ 14.5 lbf/in²]

Specific entropy, kJ/kg K

Pressure MN/m²	Sat. temp. K Liq.	Sat. temp. K Vap.	Temperature/K ... 90	100	110	120	125	130	135	140	145	150	160	170	180	190	200	250	300	Pressure MN/m²	
0.1	78.7	81.7	2.660	2.770	2.867	2.957	3.000	3.041	3.080	3.117	3.153	3.188	3.254	3.317	3.375	3.430	3.482	3.708	3.892	0.1	1 bar
0.5	96.2	98.5		2.267	2.378	2.475	2.520	2.563	2.604	2.643	2.680	2.717	2.785	2.849	2.909	2.965	3.018	3.246	3.431	0.5	
1	106.2	108.1			2.116	2.232	2.283	2.331	2.376	2.418	2.458	2.497	2.569	2.636	2.698	2.756	2.810	3.042	3.229	1	10 bar
2	118.5	119.8					1.977	2.043	2.103	2.157	2.206	2.251	2.333	2.407	2.474	2.536	2.593	2.833	3.024	2	
3	127.0	127.8				0.790	0.973	1.782	1.882	1.959	2.023	2.079	2.176	2.258	2.331	2.397	2.457	2.706	2.901	3	
4						0.739	0.888	1.074	1.443	1.752	1.854	1.928	2.045	2.140	2.221	2.292	2.356	2.614	2.812	4	
5						0.708	0.844	0.996	1.186	1.413	1.609	1.758	1.924	2.036	2.127	2.204	2.271	2.539	2.741	5	
10						0.625	0.737	0.849	0.960	1.070	1.180	1.288	1.486	1.649	1.778	1.884	1.972	2.291	2.513	10	100 bar
15						0.572	0.675	0.775	0.872	0.966	1.057	1.145	1.308	1.452	1.577	1.687	1.782	2.130	2.369	15	
20						0.532	0.630	0.723	0.812	0.898	0.981	1.059	1.206	1.339	1.457	1.563	1.657	2.015	2.265	20	
25						0.507	0.601	0.689	0.774	0.855	0.932	1.005	1.142	1.265	1.376	1.477	1.569	1.926	2.182	25	
30						0.489	0.581	0.667	0.748	0.826	0.899	0.968	1.095	1.212	1.317	1.414	1.502	1.856	2.115	30	
35													1.064	1.173	1.274	1.366	1.451	1.800	2.060	35	
40													1.040	1.144	1.240	1.329	1.412	1.754	2.013	40	
45													1.020	1.119	1.212	1.298	1.378	1.714	1.973	45	
50													1.003	1.099	1.188	1.271	1.350	1.680	1.937	50	500 bar
60															1.145	1.226	1.301	1.623	1.876	60	
70																1.187	1.261	1.575	1.825	70	
80																1.154	1.226	1.536	1.781	80	
90																		1.499	1.744	90	
100																		1.467	1.710	100	1000 bar

TRANSPORT PROPERTIES OF VARIOUS FLUIDS

In addition to giving values of the transport properties λ, μ and Pr, Tables 20–24 also list values of the thermodynamic properties v and c_p, since these are frequently required in heat transfer calculations.

Warning note:

For convenience of tabulation in Tables 20–24, the **dynamic** viscosity μ is given in g/s m instead of kg/s m ($\equiv$ N s/m^2). It should be noted, therefore, that in the calculation of viscous stress through the relation $\tau = \mu\, \partial V/\partial y$, μ must be in kg/s m if τ is to be in N/m^2, since 1 N = 1 kg m/s^2.

Similar care must be exercised when calculating the **kinematic** viscosity ν (in m^2/s) from the defining expression, $\nu = \mu/\rho = \mu v$.

TABLE 20. SATURATED WATER AND STEAM

Celsius temp., °C	Specific volume m³/kg		Isobaric specific heat capacity kJ/kg K		Thermal conductivity W/m K		Dynamic viscosity* g/s m		Prandtl number $= \mu c_p/\lambda$		Celsius temp., °C	
t	v_f	v_g	c_{p_f}	c_{p_g}	λ_f	λ_g	μ_f	μ_g	Pr_f	Pr_g	t	
0.01	0.00100	206.2	4.217	1.854	0.569	0.0173	1.755	0.0088	13.02	0.942	**0.01**	**Triple point**
10	0.00100	106.4	4.193	1.860	0.587	0.0185	1.301	0.0091	9.29	0.915	**10**	
20	0.00100	57.8	4.182	1.866	0.603	0.0191	1.002	0.0094	6.95	0.918	**20**	
30	0.00100	32.9	4.179	1.875	0.618	0.0198	0.797	0.0097	5.39	0.923	**30**	
40	0.00101	19.5	4.179	1.885	0.632	0.0204	0.651	0.0101	4.31	0.930	**40**	
50	0.00101	12.05	4.181	1.899	0.643	0.0210	0.544	0.0104	3.53	0.939	**50**	
60	0.00102	7.68	4.185	1.915	0.653	0.0217	0.462	0.0107	2.96	0.947	**60**	
70	0.00102	5.05	4.190	1.936	0.662	0.0224	0.400	0.0111	2.53	0.956	**70**	
80	0.00103	3.41	4.197	1.962	0.670	0.0231	0.350	0.0114	2.19	0.966	**80**	
90	0.00104	2.36	4.205	1.992	0.676	0.0240	0.311	0.0117	1.93	0.976	**90**	
100	0.00104	1.673	4.216	2.028	0.681	0.0249	0.278	0.0121	1.723	0.986	**100**	
125	0.00107	0.770	4.254	2.147	0.687	0.0272	0.219	0.0133	1.358	1.047	**125**	
150	0.00109	0.392	4.310	2.314	0.687	0.0300	0.180	0.0144	1.133	1.110	**150**	
175	0.00112	0.217	4.389	2.542	0.679	0.0334	0.153	0.0156	0.990	1.185	**175**	
200	0.00116	0.127	4.497	2.843	0.665	0.0375	0.133	0.0167	0.902	1.270	**200**	
225	0.00120	0.0783	4.648	3.238	0.644	0.0427	0.1182	0.0179	0.853	1.36	**225**	
250	0.00125	0.0500	4.867	3.772	0.616	0.0495	0.1065	0.0191	0.841	1.45	**250**	
275	0.00132	0.0327	5.202	4.561	0.582	0.0587	0.0972	0.0202	0.869	1.56	**275**	
300	0.00140	0.0216	5.762	5.863	0.541	0.0719	0.0897	0.0214	0.955	1.74	**300**	
325	0.00153	0.0142	6.861	8.440	0.493	0.0929	0.0790	0.0230	1.100	2.09	**325**	
350	0.00174	0.00880	10.10	17.15	0.437	0.1343	0.0648	0.0258	1.50	3.29	**350**	
360	0.00190	0.00694	14.6	25.1	0.400	0.168	0.0582	0.0275	2.11	3.89	**360**	
374.15	0.00317	0.00317	∞	∞	0.24	0.24	0.045	0.045	∞	∞	374.15	**Critical point**

* See warning note on page 31.

TABLE 21. STEAM AT ATMOSPHERIC PRESSURE

[1 atm ≈ 0.101 MN/m^2 ≈ 14.7 lbf/in^2]

Celsius temperature, °C t	Specific volume m^3/kg v	Isobaric specific heat capacity kJ/kg K c_p	Thermal conductivity W/m K λ	Dynamic viscosity* g/s m μ	Prandtl number $= \mu c_p/\lambda$ Pr	Celsius temperature, °C t
100	1.673	2.028	0.0245	0.0121	1.006	**100**
200	2.144	1.979	0.0331	0.0162	0.968	**200**
300	2.604	2.010	0.0434	0.0204	0.946	**300**
400	3.062	2.067	0.0548	0.0246	0.928	**400**
500	3.519	2.132	0.0673	0.0288	0.912	**500**
600	3.975	2.201	0.0805	0.0329	0.898	**600**
700	4.431	2.268	0.0942	0.0368	0.887	**700**
800	4.887	2.332	0.1080	0.0406	0.876	**800**

* See warning note on p. 31.

Values for water at atmospheric pressure between 0 °C and 100 °C are given with sufficient accuracy by the saturation values in Table 20.

TABLE 22. AIR AT ATMOSPHERIC PRESSURE

[1 atm ≈ 0.101 MN/m^2 ≈ 14.7 lbf/in^2]

Celsius temperature, °C t	Specific volume m^3/kg v	Isobaric specific heat capacity kJ/kg K c_p	Thermal conductivity W/m K λ	Dynamic viscosity* g/s m μ	Prandtl number $= \mu c_p/\lambda$ Pr	Celsius temperature, °C t
−100	0.488	1.01	0.016	0.012	0.75	**−100**
0	0.773	1.01	0.024	0.017	0.72	**0**
100	1.057	1.02	0.032	0.022	0.70	**100**
200	1.341	1.03	0.039	0.026	0.69	**200**
300	1.624	1.05	0.045	0.030	0.69	**300**
400	1.908	1.07	0.051	0.033	0.70	**400**
500	2.191	1.10	0.056	0.036	0.70	**500**
600	2.473	1.12	0.061	0.039	0.71	**600**
700	2.756	1.14	0.066	0.042	0.72	**700**
800	3.039	1.16	0.071	0.044	0.73	**800**

* See warning note on p. 31.

This Table may be used with reasonable accuracy for values of c_p, λ, μ and Pr of N_2, O_2 and CO

TABLE 23. CARBON DIOXIDE AT ATMOSPHERIC PRESSURE

[1 atm ≈ 0.101 MN/m² ≈ 14.7 lbf/in²]

Celsius temperature, °C t	Specific volume m³/kg v	Isobaric specific heat capacity kJ/kg K c_p	Thermal conductivity W/m K λ	Dynamic viscosity* g/s m μ	Prandtl number $= \mu c_p/\lambda$ Pr	Celsius temperature, °C t
−50	0.410	0.79	0.011	0.011	0.79	−50
0	0.506	0.83	0.015	0.014	0.78	0
100	0.694	0.92	0.022	0.018	0.75	100
200	0.881	1.00	0.030	0.022	0.73	200
300	1.068	1.06	0.038	0.026	0.72	300
400	1.255	1.11	0.046	0.029	0.71	400
500	1.442	1.16	0.053	0.032	0.70	500
600	1.628	1.20	0.061	0.035	0.69	600
700	1.814	1.23	0.069	0.038	0.68	700
800	2.000	1.25	0.078	0.041	0.67	800

* See warning note on p. 31.

TABLE 24. HYDROGEN AT ATMOSPHERIC PRESSURE

[1 atm ≈ 0.101 MN/m² ≈ 14.7 lbf/in²]

Celsius temperature, °C t	Specific volume m³/kg v	Isobaric specific heat capacity kJ/kg K c_p	Thermal conductivity W/m K λ	Dynamic viscosity* g/s m μ	Prandtl number $= \mu c_p/\lambda$ Pr	Celsius temperature, °C t
−200	2.97	10.6	0.050	0.0033	0.71	−200
−100	7.05	13.1	0.112	0.0062	0.72	−100
0	11.13	14.2	0.172	0.0084	0.69	0
100	15.19	14.5	0.220	0.0103	0.68	100
200	19.26	14.5	0.264	0.0121	0.67	200
300	23.34	14.5	0.307	0.0138	0.66	300
400	27.41	14.6	0.348	0.0154	0.65	400
500	31.48	14.7	0.387	0.0169	0.64	500
600	35.54	14.8	0.427	0.0183	0.63	600
700	39.61	14.9	0.476	0.0199	0.62	700
800	43.68	15.1	0.528	0.0211	0.61	800

* See warning note on p. 31.

APPENDIX A

DEFINITIONS OF BASIC SI UNITS

The units quoted below are the basic units of the *Système International d'Unités*. The abbreviation CGPM refers to the *Conférence Générale des Poids et Mesures.*

Length: The unit of length called the *metre* is **1 650 763.73** wavelengths *in vacuo* of the radiation corresponding to the transition between the energy levels $2p_{10}$ and $5d_5$ of the krypton-86 atom (11th CGPM, 1960).

Mass: The unit of mass called the *kilogramme* is the mass of the international prototype which is in the custody of the *Bureau International des Poids et Mesures* (BIPM) at Sèvres, near Paris, France (3rd CGPM, 1901).

Time: The unit of time called the *second* is the duration of **9 192 631 770** cycles of the radiation corresponding to the transition between the two hyperfine levels of the fundamental state of the caesium-133 atom (13th CGPM, 1967).

Thermodynamic temperature: The unit of thermodynamic temperature called the *kelvin* is the fraction **1/273.16** of the thermodynamic temperature of the triple point of water (13th CGPM, 1967).

Electric current: The unit of electric current called the *ampere* is that constant current which, if maintained in two parallel rectilinear conductors of infinite length, of negligible circular cross-section, and placed at a distance of **1** metre apart in a vacuum, would produce between these conductors a force equal to $\mathbf{2 \times 10^{-7}}$ newton per metre length (9th CGPM, 1948).

Luminous intensity: The unit of luminous intensity called the *candela* is the luminous intensity, in the perpendicular direction, of a surface of **1/600 000** metre squared of a black body at the freezing temperature of platinum under a pressure of **101 325** newtons per metre squared (13th CGPM, 1967).

Amount of substance: The *mole* is the amount of substance of a system which contains as many elementary entities as there are atoms in 0.012 kilogramme of carbon 12 (14th CGPM, 1971).

Note: When the *mole* is used, the elementary entities must be specified and may be atoms, molecules, ions, electrons, other particles, or specified groups of such particles.

APPENDIX B

DEFINITIONS OF SOME DERIVED SI UNITS

Force: The unit of force called the ***newton*** is that force which, when applied to a body having a mass of 1 kilogramme, gives it an acceleration of 1 metre per second per second. (Thus 1 N = 1 kg m/s^2.)

Pressure: The unit of pressure called the *pascal** is equal to 1 newton per square metre. (Thus 1 Pa = 1 N/m^2).

Energy: The unit of energy called the *joule* is the work done when the point of application of a force of 1 newton is displaced through a distance of 1 metre in the direction of the force. (Thus 1 J = 1 N m.)

Power: The unit of power called the ***watt*** is equal to 1 joule per second.

Electric charge: The unit of electric charge called the ***coulomb*** is the quantity of electricity transported in 1 second by a current of 1 ampere.

Electric potential: The unit of electric potential called the ***volt*** is the difference of potential between two points of a conducting wire carrying a constant current of 1 ampere, when the power dissipated between these points is equal to 1 watt.

* This name for the N/m^2 is not used in these tables.

APPENDIX C

DEFINITIONS OF SOME NON-SI METRIC UNITS

Each equation serves to define **exactly** the unit appearing on the left-hand side of the equation.

Length:

1 micron $= \mathbf{10^{-6}}\ \text{m} = \mathbf{1}\ \mu\text{m}$

1 ångström (Å) $= \mathbf{10^{-10}}\ \text{m}$

Volume:

1 litre (l) $= \mathbf{1000}\ \text{cm}^3 = \mathbf{1}\ \text{dm}^3 = \mathbf{10^{-3}}\ \text{m}^3$

Note: This is the *litre* of the 12th CGPM, 1964. It is not identical to that previously defined by the 3rd CGPM, 1901, as the volume occupied by a mass of 1 kg of pure water at its temperature of maximum density and under a pressure of 1 standard atmosphere.

1 litre (1901) $\approx 1.000\,028 \times 10^{-3}\ \text{m}^3$. (The 1901 litre is used in the definition of the UK gallon.)

Mass:

1 tonne, or metric ton (t) $= \mathbf{10^3}\ \text{kg}$

Force:

1 dyne (dyn) $= \mathbf{1}\ \text{g cm/s}^2 = \mathbf{10^{-5}}\ \text{N}$

1 kilogramme force (kgf) $= \mathbf{9.806\,65}\ \text{N}$

Note: This is that force which, when applied to a body having a mass of 1 kg, gives it an acceleration equal to the international standard acceleration of **9.806 65** m/s^2. In Germany, the *kilogramme force* is also given the name *Kilopond* (kp).

Pressure, stress:

1 bar (bar) $= \mathbf{10^5}\ \text{N/m}^2$

1 std. atmosphere (atm) $= \mathbf{1.013\,25}\ \text{bar} = \mathbf{0.101\,325}\ \text{MN/m}^2$

1 tech. atmosphere (at) $= \mathbf{1}\ \text{kgf/cm}^2$

$= \mathbf{0.980\,665}\ \text{bar} = \mathbf{0.098\,0665}\ \text{MN/m}^2$

Note: This unit is also sometimes given the unit symbol *ata*.

1 torr $= \mathbf{1/760}\ \text{atm} \approx 133.0\ \text{N/m}^2$

$\approx$ 1 mmHg to within 1 part in 7 million

1 mmHg $= \mathbf{13.5951} \times \mathbf{9.806\,65}\ \text{N/m}^2$

$\approx 133.0\ \text{N/m}^2$

Note: This is the pressure that would be exerted by a **1** mm column of mercury of density **13.5951** g/cm^3 under a gravitational acceleration equal to the international standard acceleration of **9.806 65** m/s^2.

Energy:

1 erg $= \mathbf{1}\ \text{dyn cm} = \mathbf{10^{-7}}\ \text{N m} = \mathbf{10^{-7}}\ \text{J}$

1 calorie (cal) $= \mathbf{4.1868}\ \text{J}$

Note: This is the *International Table calorie*, defined thus by the Fifth International Conference on the Properties of Steam, 1956.

1 thermochemical calorie $= \mathbf{4.184}\ \text{J}$

Dynamic viscosity: 1 poise (P) $= \mathbf{1}\ \text{g/cm s} = \mathbf{1}\ \text{dyn s/cm}^2$

$= \mathbf{0.1}\ \text{kg/m s} = \mathbf{0.1}\ \text{N s/m}^2$

Kinematic viscosity: 1 stokes (St) $= \mathbf{1}\ \text{cm}^2/\text{s} = \mathbf{10^{-4}}\ \text{m}^2/\text{s}$

APPENDIX D

BRITISH UNITS—DEFINITIONS AND CONVERSION FACTORS

DEFINITIONS OF SOME BASIC UNITS

Each equation serves to define **exactly** the unit appearing on the left-hand side of the equation.

Length: 1 yard (yd) = **0.9144** m

Mass: 1 pound (lb) = **0.453 592 37** kg

Force: 1 poundal (pdl) = **1** lb ft/s²

1 pound force (lbf) = $\frac{9.806\,65}{0.3048}$ pdl (≈ 32.2 pdl)

Note: This is the force which, when applied to a body having a mass of 1 lb, gives it an acceleration equal to the international standard acceleration of **9.806 65** m/s².

Specific energy: 1 Btu/lb = **5/9** cal/g

Note: This equation serves to define the *British Thermal Unit* (Btu). The calorie is here the *International Table calorie* defined in Appendix C.

CONVERSION FACTORS FOR MECHANICAL UNITS

Quantity	British unit	Conversion factor		SI unit
Length	1 inch (in)	$= \frac{0.9144 \times 100}{36}$	= **2.54**	cm
	1 foot (ft)	$= \frac{0.9144}{3}$	= **0.3048**	m
	1 mile (mile)	$= \frac{0.9144 \times 1760}{1000}$	≈ 1.61	km
Mass	1 ounce (oz)	$= \frac{0.453\,592\,37 \times 1000}{16}$	≈ 28.35	g
	1 pound (lb)	= 0.453 592 37	≈ 0.4536	kg
	1 hundredweight (cwt)	= 0.453 592 37 × 112	≈ 50.8	kg
	1 ton (ton)	= 0.453 592 37 × 2240	≈ 1016	kg
	(1 US short ton	= 0.453 592 37 × 2000	≈ 907	kg)
Force	1 poundal (pdl)	= 0.453 592 37 × 0.3048	≈ 0.1383	N
	1 ounce force (ozf)	$= \frac{0.453\,592\,37 \times 9.806\,65}{16}$	≈ 0.278	N
	1 pound force (lbf)	= 0.453 592 37 × 9.806 65	≈ 4.45	N
	1 ton force (tonf)	= 0.453 592 37 × 9.806 65 × 2240	≈ 9.96	kN
Volume	1 UK gallon (gal)	$= \frac{8.134\,783}{0.997\,642} \times \frac{4535.9237}{8.136 \times 10^6} \times 1.000\,028$	≈ 4.546	dm³

Definition: The UK gallon (*imperial gallon*) is the space occupied by **10** pounds *weight* of distilled water of density **0.998 859** gramme per millilitre weighed in air of density **0.001 217** gramme per millilitre against weights of density **8.136** grammes per millilitre. (The *litre* is here the 1901 litre defined in Appendix C.)

(This is very nearly equal to the volume that would be occupied by 10 lb *mass* of water of density 62.3 lb/ft³.)

	(1 US gallon	$= 231 \times (2.54)^3 \times 10^{-6}$	≈ 3.785	dm³)
Specific volume	1 ft³/lb	$= \dfrac{(0.3048)^3}{0.453\ 592\ 37} \times 10^3$	≈ 62.43	dm³/kg
Pressure, stress	1 tonf/in²	$= \dfrac{0.453\ 592\ 37 \times 9.806\ 65}{10^6}\left(\dfrac{12}{0.3048}\right)^2 \times 2240$	≈ 15.44	MN/m²
	1 lbf/in²	$= \dfrac{0.453\ 592\ 37 \times 9.806\ 65}{10^3}\left(\dfrac{12}{0.3048}\right)^2$	≈ 6.895	kN/m²
	1 inHg*	$= 25.4 \times 13.5951 \times 9.806\ 65 \times 10^{-3}$	≈ 3.39	kN/m²
	1 ftH₂O*	$= 0.3048 \times 9.806\ 65$	≈ 2.99	kN/m²

* *Note:* These are equal respectively to the pressures that would be exerted by a 1 in column of mercury of density **13.5951** g/cm³, and by a 1 ft column of water of density **1** g/cm³, under the international standard acceleration of **9.806 65** m/s².

Dynamic viscosity	1 lb/ft s = 1 pdl s/ft²	$= \dfrac{0.453\ 592\ 37}{0.3048}$	≈ 1.488	kg/m s N s/m²
Kinematic viscosity	1 ft²/s	$= (0.3048)^2$	≈ 0.0929	m²/s
Energy	1 ft lbf	$= 0.3048 \times 0.453\ 592\ 37 \times 9.806\ 65$	≈ 1.356	J
Power	1 horsepower (hp)	$= 550 \times 0.3048 \times 0.453\ 592\ 37 \times 9.806\ 65$	≈ 746	W
Specific fuel consumption	1 lb/hp h	$= \dfrac{10^6}{550 \times 0.3048 \times 9.806\ 65 \times 3600}$	≈ 0.169	kg/MJ

CONVERSION FACTORS FOR THERMAL UNITS

Energy	1 Btu	$= \dfrac{4.1868 \times 0.453\ 592\ 37}{1.8}$	≈ 1.055	kJ
	1 therm (= 10⁵ Btu)	$= \dfrac{4.1868 \times 0.453\ 592\ 37}{1.8} \times 10^2$	≈ 105.5	MJ
Specific energy	1 Btu/lb	$= \dfrac{4.1868}{1.8}$	$= 2.326$	kJ/kg
Sp. heat-capacity **Sp. entropy**	1 Btu/lb R	$= 4.1868$	≈ 4.19	kJ/kg K
Gas constant	1 ft lbf/lb R	$= 0.3048 \times 9.806\ 65 \times 1.8$	≈ 5.38	J/kg K
Thermal conductivity	1 Btu/h ft R	$= \dfrac{4.1868 \times 0.453\ 592\ 37}{3.6 \times 0.3048}$	≈ 1.73	W/m K
Heat transfer coefficient	1 Btu/h ft² R	$= \dfrac{4.1868 \times 0.453\ 592\ 37}{3.6 \times (0.3048)^2}$	≈ 5.68	W/m² K

Note: R is here the *rankine* unit of *thermodynamic temperature*, here defined in terms of the *kelvin* by the relation 1 R = (1/1.8) K.

APPENDIX E

TEMPERATURE

Thermodynamic temperature

In constructing *thermodynamically consistent* tables of thermodynamic properties, use has to be made of equations such as $Tds = dh - v\,dp$, in which the symbol T refers to the *thermodynamic temperature* defined by the equation

$$\frac{T_1}{T_2} = \frac{Q_1}{Q_2},$$

where Q_1 and Q_2 are respectively the quantities of heat received by and rejected from a cyclic heat power plant operating ***reversibly*** between two thermal-energy reservoirs at temperatures T_1 and T_2.

Zero thermodynamic temperature is that temperature to which, with T_1 at a fixed positive value, T_2 tends as Q_2 tends to zero. It is unattainable in practice, but this ***absolute zero of temperature*** nevertheless constitutes a definite, fixed level of temperature. (The unattainability of this absolute zero of temperature might appear to result from the fact that, if Q_2 were zero, the cyclic heat power plant would constitute a ***perpetual motion machine of the second kind***, so that the Second Law would be contravened. The point is more subtle than this, however, and is discussed by A. B. Pippard in Chapter 5 of *Elements of Classical Thermodynamics*, Cambridge University Press, 1957.)

The kelvin unit of thermodynamic temperature

With **zero** thermodynamic temperature defined, it is only necessary to assign an arbitrary number of units to some other *temperature level* in order exactly to define the *unit* of thermodynamic temperature.

The kelvin unit of thermodynamic temperature, defined in 1954 and redefined in 1967, when it was given the title ***kelvin*** and the *unit symbol* K, is that obtained by assigning to the temperature level at the ***triple point*** of water a value of **273.16** kelvins, namely **273.16** K. The precise definition of this unit is given in Appendix A.

Note: Unfortunately, thermodynamic temperatures are frequently written as T °K instead of, as here, T K.

Celsius temperature

By virtue of long-established habit, it is convenient in practice to use a truncated thermodynamic temperature called the *Celsius temperature*, defined by the relation

$$t = T - 273.15,$$

where t *Celsius* (written symbolically as t °C) is the Celsius temperature at a thermodynamic temperature of T kelvins (namely T K).

The convenience of this practice arises from the fact that the Celsius temperature at the *ice point* (the freezing point of air-saturated water at 1 atm) is then very nearly, though not exactly, 0 °C.

Note: A thermodynamic temperature expressed in kelvins is commonly described as the *absolute* temperature, in order to distinguish it from the *Celsius temperature*. That usage is followed in the warning note given on those tables in which temperatures are listed in kelvins.

APPENDIX F

CONVERSION FACTORS FOR TEMPERATURE

In the British system of units, the unit of thermodynamic temperature, here given the name *rankine* and the unit symbol R, is defined in terms of the kelvin by the relation

$$1\ \mathrm{R} = (1/1.8)\ \mathrm{K},$$

so that at a thermodynamic temperature of T_k kelvins (namely T_k K) the thermodynamic temperature is T_r rankines (namely T_r R), where

$$T_r = 1.8\, T_k.$$

In the same way as it is convenient in practice to use a truncated thermodynamic temperature called the *Celsius temperature*, it is also convenient to use its counterpart in the British system of units, the *Fahrenheit temperature*, defined by the relation

$$t_F = T_r - 459.67,$$

where t_F *Fahrenheit* (written symbolically as t_F °F) is the Fahrenheit temperature at a thermodynamic temperature of T_r rankines (namely T_r R). In this way, the Fahrenheit temperature at the *ice point* is very nearly, though not exactly, 32 °F.

At a Celsius temperature of t_C °C, the Fahrenheit temperature, t_F °F, is given by the relation

$$t_F = 1.8\, t_C + 32,$$

or, more conveniently,

$$(t_F + 40) = 1.8\,(t_C + 40).$$

Note: The nomenclature and symbology used in this Appendix in relation to British units have not been standardised. Thermodynamic (absolute) temperatures are frequently written as T °R instead of, as here, T R.

Temperature differences are expressed in kelvins or rankines, for which the unit symbols are respectively K and R, so that the use of such symbols for temperature difference as *degK*, *degC*, *degR*, and *degF* is avoided.

APPENDIX G

PRINCIPAL SOURCES OF DATA

Maxwell, T. B. *Data Book on Hydrocarbons*. Van Nostrand, New York, 1950.

Keenan, J. H. and Kaye, J. *Gas Tables*. Wiley, New York, 1948.

Janaf Thermochemical Tables. The Dow Chemical Co., Midland, Michigan, 1965/1966.

The 1967 IFC Formulation for Industrial Use (*A formulation of the thermodynamic properties of ordinary water substance*). International Formulation Committee of the 6th ICPS, IFC Secretariat, Verein Deutscher Ingenieure, Düsseldorf, 1967. (Reproduced in *1967 Steam Tables*, E.R.A./Arnold, London, 1967.)

McHarness, R. C., Eiseman, B. J. and Martin, J. J., *The New Thermodynamic Properties of 'Freon-12'*, Refrigerating Engineering, Vol. 63, No. 9, p. 31, September 1955 (their equations having been put into simplified form by the author for computation of Table 13 and for the plotting of the pressure–enthalpy diagram for Refrigerant-12).

Properties of Commonly-used Refrigerants. Air Conditioning and Refrigerating Machinery Association, Inc., Washington, D.C., 1946.

Теплофизические свойства двуокиси углерода, Вукалович, М. П. и Алтунин, В. В., Атомиздат, Москва, 1965. (*Thermophysical Properties of Carbon Dioxide*, Vukalovich, M. P. and Altunin, V. V., Atomizdat, Moscow, 1965. Translation from the Russian edited by D. S. Gaunt, Collett's (Publishers) Ltd, London, 1968.)

Thermodynamic Functions of Gases. Din, F., ed. Butterworths, London, 1962.

The 1967 Steam Tables. Published for the Electrical Research Association by Ed. Arnold, London, 1967. (Used in preparing Tables 20 and 21.)

Hilsenrath, J. et al., *Tables of Thermodynamic and Transport Properties* (N.B.S. Circular 564), Pergamon Press, Oxford, 1960.

Справочник по теплофизическим свойствам газов и жидкостей, Варгафтик, Н. Б., Государственное Издатепьство Физико-Математической Литературы, Москва, 1963. (*Reference Book of Thermophysical Properties of Gases and Liquids*. Vargaftik, N. B., State Publishing House of Physical and Mathematical Literature, Moscow, 1963.)

Conversion Factors and Tables. British Standard 350: Part 1: 1959 (and Amendments No. 1 and 2, 1963), British Standards Institution, London.